AUSTRALIAN

Bush Flower
—Essences—

A U S T R A L I A N

Bush Flower
— Essences —

IAN WHITE

BANTAM BOOKS
SYDNEY • AUCKLAND • TORONTO • NEW YORK • LONDON

In memory of
Paracelsus and Edward Bach, the modern founders of
flower essences

AUSTRALIAN BUSH FLOWER ESSENCES

A Bantam Book

Printing history
Bantam edition first published 1991
Reprinted in 1991, 1992, 1993 and 1994

National Library of Australia
Cataloguing-in-Publication data

White, Ian.
Australian bush flower essences.

Bibliography.
Includes index.
ISBN 0 947189 75 0.

1. Medicinal plants – Australia. 2. Wild flowers –
Australia – Therapeutic use. 3. Essences and essential
oils. I. Title.

581.6340994

Bantam Books are published by

Transworld Publishers (Aust.) Pty Limited
15–25 Helles Avenue, Moorebank, NSW 2170

Transworld Publishers (NZ) Limited
3 William Pickering Drive, Albany, Auckland

Transworld Publishers (UK) Limited
61–63 Uxbridge Road, Ealing, London W5 5SA

Bantam Doubleday Dell Publishing Group Inc.
1540 Broadway, New York, New York 10036

Typeset by Midland Typesetters, Maryborough, Victoria
Printed in Singapore by Kyodo Printing Co. Ltd.

Photographs: Ian White
Illustrations: Kristin White
Designed by Trevor Hood
Edited by Jacquelin Hochmuth

10 9 8 7 6 5

—Contents—

—Foreword—

Ian and Kristin White are to me the personification of the aroma of the Brown Boronia.

The perfume of their spirit is so gentle, fragile and exquisite that they give me real hope for the future.

To be in the bush with them as I have, to feel their reverence and respect for all things natural, one cannot but feel that every step they take is a prayer. Their spirits have been encompassed, like the Aborigines of long ago, by the energies of this great continent, Australia. And, like me, they like nothing better than to meander through our ever-changing landscape, observing, sampling, feeling the vibrations—and learning.

I commend this book, which is given to you in love.

BURNUM BURNUM

—— Acknowledgements ——

Firstly to Spirit. Thank you for all your unseen help and guidance. I am also very pleased to have the opportunity to thank some very special people whose assistance was invaluable to me in writing this book.

Firstly, three wonderful friends:

Malcolm Cohan, for his continual encouragement and for helping me come to terms with my computer and unravel its mysterious workings. For his inspiration and support at times when I was feeling very depressed about the whole project—times when complete files would suddenly disappear into the Never-Never. For always being on hand to help me pick up the pieces and patiently going over all the rules I should have been following . . . again.

Also, Mal's wife, Susan Hayward, for her support and for graciously allowing me to use so much of her research for the quotes in her books, *A Guide for the Advanced Soul*, *Begin It Now* and, in collaboration with Malcolm, *A Bag of Jewels*.

And Mark Brodbeck, for constantly being there to help my staff and me become computer literate and for being a real whiz when Max, the computer, did inexplicable things.

To all three, who had such great faith in me and the project and gave of themselves and their time so freely, *thank you very much*—it's great to have such good friends.

To Budget Rent A Car, special thanks for the support and expertise we have always received during our trips to the Outback.

Of the hordes of others who helped me, I'd like to single out and thank the following:

My mother, for all her help in many ways over many years; my father-in-law, John Coburn; David Phillips, the man who introduced me to numerology and metaphysics and taught me those subjects; Glynn Braddy, another mentor who gave me profound and wonderful insights into healing and metaphysics; Nevill Drury, for the genesis of the idea for a book—even Darwin would have been proud of the evolution of the original concept; Barbara McGregor, for her crucial initial encouragement; Jeanette Bakker and Owen Davidson, for confirmation and further understanding of the healing properties of the Australian Bush Flower Essences; and finally, and certainly most importantly, my wife, Kristin, for everything and then a bit more—and our new daughter, Grace, for the energy she has given to this project.

To everyone else who assisted, thank you.

IAN WHITE
1990

—Introduction—

As a small boy I grew up in the Australian bush in an area north-west of Sydney called Terrey Hills. I would often accompany my grandmother on long walks through the bush as she gathered medicinal herbs. From her profound understanding she would point out to me the different plants and flowers and explain their various properties. Her respect for nature rubbed off on me, as did her appreciation of the Australian bush with its uniqueness and tremendous power.

Geologically, Australia is the oldest of all the continents, and there is certainly an old wisdom that can be felt when one travels through the land, especially in the outback—a wisdom coupled with tremendous strength.

Australian plants have evolved in isolation over the last forty-five million years. One hundred and eighty million years ago the world consisted of a single land mass known as Pangaea. This then split in two, with Laurasia constituting the northern continents, and Gondwanaland the southern. The latter was made up of Australasia, Antarctica, India, Africa and South America. India and Africa drifted away more than sixty-five million years ago leaving Antarctica sandwiched between Australasia and South America. These three continents themselves began to split apart forty-five million years ago, with Australia drifting to a less temperate zone.

This new warmer, drier climate resulted in the extinction of many plants, while others were able to adopt new characteristics that allowed them to survive, leading to the development of many rare and unique genera and species. Today Australia is regarded as possessing, along with Brazil, the highest number of flowering plants in the world.

Australian plants are very distinctive as many have prickly foliage, an adaptation to the arid areas in which they grow. Those with hardened leaves are known as sclerophylls. Also there is a very peculiar scent that emanates from the Australian bush. Sailors and other seafarers can often smell the bush before catching sight of land.

Even as a small boy I was aware that there was something quite remarkable about the Australian bush. It has a wonderful inherent beauty and strength, and there are always a great many plants in flower no matter what the season. You can go through the bush in the middle of winter and still see a blaze of colour from the myriad flowers, mingled with the stark and also subtle hues of their surroundings. People, especially those from the Northern Hemisphere, are very surprised to see so many flowers all through the year as opposed to the predominantly spring and summer flowering on the other continents. I now know that the colours I noticed in those days at Terrey Hills—the striking reds and purples—predominate throughout the country. Metaphysically, red symbolises raw physical energy, and purple, higher learning. It is as if these flower colours represent the spiritual wisdom present in Australia and the strong new vitality that is now here too.

Australia has always had a very wise old energy, yet at the moment there is also a tremendous vitality in this country. This metaphysical energy manifests in the Australian Bush Flower Essences, and both Australia and these remedies have major roles to play in the unfolding of the New Age consciousness. Australia is becoming the centre of New Age inspiration and learning, and the bush essences will be instrumental in the development of emotional and spiritual growth. The same vitality that was evident in ancient Egypt, Greece 2500 years ago, when Hippocrates, Plato and Pythagoras were alive, and that was later found in Rome, is now here in Australia.

There is a theory that this energy is present in only one country at any one point in time, and that over the centuries it has been continually shifting from place to place. A study of history will show the course of this flow in the rise of great nations or empires. From the ancient Egyptian, Hebrew and Greek civilisations and the Chinese dynasties; to the Roman Empire; initially to Italy and then to France during the Renaissance; Japan, Britain and Germany; and more recently it was centred in the west coast of America and in Hawaii.

This energy, combined with the strength and wisdom of Australia, is now manifest in her flora and encapsulated in the flower essences made from those plants. Consequently, the Australian Bush Flower Essences are and will continue to be used in other countries without losing any of their healing qualities or potency—they will become universal remedies.

Generally, it is much better to use local plants for healing the inhabitants of any area, as those plants share the same environment and are in harmony with the area and the people living there. Yet in many countries the local people are not making up their own essences from the plants growing among them. But as the Australian essences have universal application today, they are beneficial to people everywhere.

Another factor in the emergence and potency of the Australian Bush Flower Essences is that the plants from which they are made enjoy a clean environment. Unlike other large land masses, Australia is relatively unaffected by the scourges of nuclear and chemical pollution.

At this point it seems appropriate to relate some of the events that led to the development of the bush essences. A particular theme has run through my family for five generations. My grandmother was a herbalist, as was her grandmother, my great-great-grandmother, who practised in New Zealand. Her daughter, my great-grandmother, was also a herbalist, initially in New Zealand and later in Australia. My grandmother specialised in treating children. Her son, my father, was a pharmacist, though he prescribed and dispensed many herbal medicines. Customers coming in with various complaints would often end up with a herbal mixture rather than the standard pharmaceutical product.

My grandmother and great-grandmother worked with many native plants and built up a great deal of information on them. Unfortunately, much of this knowledge was never written down and was lost when my grandmother died. She had cancer for many years, possibly due to the very strong cigarettes that she smoked, but with her herbs she was able to keep the cancer under control. When she travelled to Europe, the luggage containing all her herbs went astray and it was three months before she was able to renew her supply. The cancer spread during that time and, weakened by the stress of travelling, she died soon after returning home.

Even with that lineage, I had no initial intention of taking up herbal medicine or naturopathy but enrolled at university to study science and psychology. During one of the long summer vacations I went travelling in India and, like many other travellers to that country, I came back very ill with dysentery. I had left strong and healthy and returned the exact opposite—very tired and lethargic, with skinny arms and legs and a pot belly. My family attempted to build up my strength by feeding me three hot meals a day. However, their attempt failed as I didn't have the energy to burn up all that food. I still had skinny arms and legs but my pot belly had grown larger.

At that point I decided to take responsibility for my own health and moved into a household where there was a great deal of interest in alternative medicine. An acupuncturist and a chiropractor, among others, lived there. During that time I became reacquainted with many concepts of alternative medicine, and I decided to combine psychology and natural therapies in order to complete my university degree. As in many other tertiary institutions teaching psychology, only a very small proportion of the lecturers—two out of twenty-four—was involved in what could be called humanistic psychology. The more I learned of the philosophy of natural therapies, the more I realised that it explains the workings of the human psyche much more comprehensively than academic psychology.

It was at this point that I became, for the first time in my life, very clear about what I wanted to do, so I enrolled at the New South Wales College of Natural Therapies in the then only full-time course available in naturopathy. I graduated and have been practising for over ten years. While practising, I taught the use of English flower essences and also prescribed them with good results. Even so, my feelings for the Australian bush and the immense healing potential of Australian plants never left me. I was constantly amazed that those plants were not being used. It seemed such a great waste. Little did I realise what was to follow.

So much for my academic background. The other area relevant to the development of the bush essences is what could be loosely called my metaphysical background.

On returning from India with my health and emotions in tatters, I explored various forms of yoga and meditation, including Zen meditation. Eventually I was led back to my roots and introduced to various spiritual circles in which prayer, meditation and direct healing energy predominated.

Then a major turning point occurred in my life. A very dear friend, Jim Vicary—only in his early thirties—told me that he had been diagnosed as having bowel cancer and that he was returning home to Brisbane for surgery. He asked if my wife Kristin and I would hold a meditation healing circle with our mutual friends the night before his surgery in order to direct healing energy to him. During the operation the doctors found that the cancer had spread to Jim's liver, and they said that there was nothing they could do and gave him only a few months to live.

Every Monday night from then on this same group of friends came to our house and we held a healing circle. After a couple of months the guidance on the bush essences started coming through. While meditating, I was shown a picture of a certain plant and the best place in which to make up its essence. I was also given an understanding of its healing properties. If I was unfamiliar with a flower, its name would appear beneath it.

This was a very exciting time, as the fact that Australian plants have healing properties was being confirmed for me—something I had known deep inside since I was a small child. There was no doubting the authenticity of the results. Once I was told to go to a well-known headland to make up the essence of a particular flower. I had lived nearby years before and had never known the plants to be prolific in that area. Yet this time the whole headland was a mass of these flowers.

There was also a wonderful synchronicity associated with the bush essence remedies. Often I would go out in the morning to make up an essence and later that day I would see patients who may never have consulted me before but who would invariably present with problems that were indicative of, and could be greatly helped by, that particular essence.

With an established practice, I was in a perfect position to observe the results of using the bush essences on my patients. Certainly their comments and their improved physical condition showed that the essences were far more powerful and faster acting than any other remedies I had ever prescribed. I joke that I am a lapsed homoeopath, as homoeopathy had previously been a large part of my practice. The bush essences have replaced that modality to a great extent.

As a result, a number of other practitioners, including those working with and sharing my keen interest in flower essences, also became very excited by my results and began prescribing the bush essences.

Through our observation of the effects of the essences, we were able to verify the information that was being channelled to me during meditation. Although I was keen to check the validity of that information, I never resorted to double-blind studies, as I always wanted to do all that I could to help my patients heal themselves and I saw the essences as being very powerful for that purpose. However, those studies are now being done by others.

In order to check the accuracy of my channellings further before publishing any information on the essences, I and others rigorously challenged my understanding of the remedies with kinesiology, Kirlian photography, and medical electronic diagnostic equipment such as Vega and Morey machines. As well, I sought the help of a number of trance and conscious mediums whom I respected and trusted. I compared my information with the material that came through them. Again, and with only one small addition to one out of all fifty remedies, there was complete verification.

Today there are fifty Bush Flower Essences which have been made up by myself and Kristin from plants all over Australia. They are being used by practitioners and by those wanting to use the essences on themselves, and/or their friends and families, throughout many countries.

For me, the bush essences have been a wonderful journey. I have learned photography and botany, I have written several books, travelled to marvellous places and made wonderful friends.

The bush essences themselves have a tremendously important role to play. They are powerful catalysts for helping people heal themselves. The essences allow people to turn inwards and understand their own life plan, their own life purpose and direction. They also give people the courage and confidence to follow that plan. Illness, disease and emotional problems are only indicators that we have strayed off our individual path. The essences, as well as helping us return to that path, can assist us to work through and resolve our problems and imbalances. They can also help to give us an understanding of why these difficulties came about in the first place and what needs to be done to clear them, by unleashing the positive qualities inherent in us.

I feel that we are approaching a point in human history where we are at the crossroads. We now have the opportunity to make some really significant advances in the quality of our lives by improving the spiritual, physical and emotional aspects of the human condition. I see the bush essences as being one of the major keys that will enable us to advance to the point where the individual, the society and the whole planet benefits.

The following is a channelled message received from the Bush Essences. "We wish you to think of us as a 'body', so to speak. We have been brought forward and given to mankind through a messenger from our heavenly Father. We have come to unite heaven and earth. For we are the fruits of earth and you dear people are the fruits of heaven. We are united in your hearts. We are grateful to be with you—to be able to work through you to help mankind."

The History and Purpose of Flower —Essences—

Flower essences have a very long history spanning many cultures. Even before Christ's time they were used for health and healing. In fact, a number of books claim that flower essences were widely used in the ancient and esoterically known civilisations of Lemuria and Atlantis.

The Australian Aborigines obtained the beneficial effects of a flower essence by eating the whole flower. The essence, in the form of dew made potent by the sun, would thus be consumed with the flower. At the same time, the Aborigines benefited from the nutritive properties of the flower. Often they didn't distinguish between the plant and the flower and simply ate both. Or sometimes they ate the flower just for its taste, especially if it was rich in honey. Moreover, if a flower was inedible, they would sit in a clump of flowers to absorb the healing vibration of the flower.

The use of a flower for the purpose of healing was always seen by Aborigines as a pleasurable rite. They certainly knew of many flowers that could be used for resolving specific emotional imbalances.

A number of other cultures, including the Egyptian, Malay and African, used flowers to treat emotional states and imbalances. Though there is

European folklore on the healing power of flowers that dates back to at least medieval times, the earliest recorded use of flower essences occurred in the sixteenth century when the great healer and mystic Paracelsus collected the dew from flowers to treat his patients' emotional imbalances.

For the ancient herbalist, an understanding of the healing properties of plants was based on the Doctrine of Signatures, whereby some peculiarity of a plant, such as its shape, growth, colour, scent, or taste, indicated its healing properties. For example, eyebright, a blue flower with a yellow centre, suggests the human eye and was used to treat tired eyes. The skullcap flower resembles the shape of the human skull and was used for headaches and insomnia. Nettles are good for poor circulation, and the bark of the willow eases rheumatism, which becomes worse in damp weather—the tree grows in wet places. Arrach is a foul-smelling plant which is used for foul ulcers. Flowers used to treat jaundice, such as dandelion, agrimony and celandine, are yellow in colour.

It would appear that the knowledge of the healing properties of flower essences was largely lost during the last few centuries, especially in the Western world, but to some degree even among the Aborigines.

However, the modern pioneer of flower essences, Dr Edward Bach (1886-1936), brought flower remedies back into use. The timing of Bach's life and work was perfect. His understanding of plants and their properties came into being just long enough before our time of greater awareness to enable people to become familiar with the concept of flower essences and their unique form of healing. This, apart from the wonderful healing effects of these early essences, remains one of if not *the* greatest legacy of Bach's work.

Initially, very little research into flower essences was carried out after Bach's death. England was, and still is, a country of tradition, so a certain reverence was shown towards the Bach flower remedies and it was felt that they were complete in themselves. But today, sixty years later, we know that those pioneer remedies did not address such areas as sexuality, communication, learning skills, creativity and spirituality, which are so relevant to modern society.

However, in the last ten years or so a great deal of research has been done in various parts of the world, for the time is now ripe for the elevation of flower essences to their destined role as one of the most important of the major systems of healing.

This resurgence over the last decade or so is interesting, for throughout history a phenomenon has often occurred in the world of ideas, whereby a number of people, often unaware of each other, either have an identical insight or else embark upon a similar project at the same time. In art there was the example of Picasso and Braque who, independently of one another, both painted Cubist pictures which heralded in modern abstract art. While in science there was the almost simultaneous discovery of the composition of water by Watt, Cavendish and Lavoisier; the race for the discovery of the genetic code by Watson, Crick and Linus Pauling; as well as the parallel investigations of Darwin and Wallace into the Theory of Evolution. These are only a few of the numerous instances of the phenomenon.

One explanation puts considerable emphasis on external forces, particularly those of a socio-economic nature, rather than on the thoughts and actions of the people involved.

I tend to see it from a metaphysical point of view. As far as current thought is concerned, I feel that time is offering us an amazing opportunity to burn the dross from our personalities and souls, which will allow us to attain unknown heights in the quality of our own lives and in the interactions and relationships we have with one another. As we approach this period of change, the skills and tools that are needed for this transition are manifested through the insights and inspirations of those who happen to be able to tap into the universal consciousness or directly into Spirit. For it is from those sources that we are given the new understanding or wisdom necessary to shift us to the next point of awareness.

Innovations like the Australian Bush Flower Essences are especially relevant to the modern world. In the last sixty years numerous changes have occurred in our awareness of ourselves and our world, and the essences have evolved as a form of healing to help us keep pace with the changes. Nature always has something to offer people as they evolve. The threat of global annihilation through nuclear war, the environmental crisis, the rate at which we receive information and rapid technological change have all required major shifts in our consciousness, and there are now essences to help us cope with these shifts.

Anyone can make up a flower essence as the technique is basically very simple. The skill or challenge lies in determining the healing qualities of the flower. Although every person has the potential to determine the healing properties of a specific plant or essence, some have been blessed with a special ability for doing so.

This gift, as I see it from my own perspective and experience, is like a funnel whose brim can be widened immeasurably by meditation and other intuitive practices. This funnel allows a great number of devas, angels, guides and helpers from the Spirit World to communicate collectively with an individual working with that gift. Usually one of the person's spirit guides integrates all the information into a uniform whole and channels it down—through the neck of the funnel, so to speak—as knowledge, a feeling, or a vision. When the brim is widened, the sphere of influence draws in not only more helpers, but those of a more highly evolved nature as well. The channellings from the latter flow more easily and are of a higher quality.

It is a tremendously exciting and at the same time humbling experience to work with this gift; to realise that for many years a ceaseless and tireless effort has been made by many souls in Spirit to guide one in developing the necessary skills to use this gift; and to try to comprehend the scope, potential and significance of the channellings.

Certainly Bach himself felt he should take little credit for his work as he was merely an instrument of God. Douglas Baker, in his book *Esoteric Healing*, claimed that the poet, psychic and mystic Robert Browning was, from the Other Side, directing and guiding Bach in his discoveries.

The role and relevance of flower essences in modern healing has already been discussed, and I feel it is for the same reasons that a number of people have suddenly begun the work of channelling and making up these important new essences.

One thing that has always been apparent to me is the magic at work in the making up of a bush essence, in the form of help and guidance from Spirit. The magic has always been far too obvious to be dismissed (see "The Magic of Making Bush Essences"). And this is indeed confirmation that the Spirit World (or Christ Consciousness or the Light) certainly wants us to have the essences, and that a very important role exists for them at this point in time.

There is, however, a conservative body of thought claiming that Bach's English flower remedies are still adequate for today's needs. In fact, some argue that his system was and always will be sufficient for all of humanity's requirements. To my mind, the help and magic that are so evident in the making up of these new bush essence remedies is testimony that this is not the case. For, otherwise, why would there be guidance and direction from both this world and that of Spirit enabling us to make up these essences successfully? And surely it is great arrogance to think that there will only ever be thirty-five English, one Tibetan, one Swiss and one Italian flower that have healing powers, that we could never discover flowering plants with equivalent or more potent healing powers than those thirty-eight or that there could not be flowers whose essences address other emotional imbalances.

Anna Flora, one of the first people to be involved in the making of flower essences after Bach's death, is one of many people who have had communication with Dr Bach, to express his support and to let it be known that he is guiding and inspiring people today to develop new essences. On numerous occasions I have received similar messages of great encouragement from Spirit through a variety of accurate and authentic sources.

Australian remedies work quickly and deeply, even on someone who is not living in Australia—provided that person feels comfortable with their concept of Australia. If they feel very negative towards Australia, then the bush essence remedies won't be as effective for them. However, if their concept of Australia is neutral or favourable, and if they have done some work on themselves to develop emotionally and spiritually, then the Australian Bush Flower Essences will be unsurpassed for the speed and effectiveness of their action. Remember, too, that the bush essences address the psychological, spiritual and physical states relevant to today.

In June 1987 we received a newsletter from the World Network Centre for the Harmonic Convergence in Glastonbury, England. Its front page claimed what we already knew deep inside . . . that flower medicine would become the new and major form of healing after the Harmonic Convergence. This was yet another confirmation of the relevance of our work with the bush essences.

The power of the bush essences and the results they constantly produce are astounding. They act as catalysts to help resolve a vast range of negative emotional states and develop intuitive abilities. They heal by helping to bring a person into emotional, spiritual and mental harmony. The rationale behind their healing capacity is based on the timeless wisdom that when emotional balance is restored, true healing occurs. Most physical illness is the end result of emotional imbalance.

But the main purpose of the essences is to help people get in touch with their Higher Self—their own intuitive centre that knows their life's purpose. It is now time for people to learn, stand up for and then follow what they want and need to do. To know that they have the power to make positive changes not only for themselves but also for this planet, and that they can really make a difference. The more people use the essences, the more they experience clarity and quality in their lives. They also begin to shape and control the economic and social changes occurring around them.

The bush essences have extraordinary healing powers and an extremely important role to play in helping to heal the planet and raise the level of awareness of those living on it. To quote Anna Flora: "The essences help make humans more like angels."

How
Bush Essences
—Work—

The concept of healing that was shared by such great healers as Hippocrates, Paracelsus, Hahnemann, Bach and Steiner was a simple one. They all believed that good health was the result of emotional, spiritual and mental harmony and found that when they treated their patients' psychological imbalances, their diseases were cured.

Disease itself is the physical manifestation of emotional imbalances which predominantly occur when people are not in touch with their Higher Selves. Yet disease is not something to be feared or worried about. Rather, it can be seen as an indication that something in a person's life is out of balance. Consequently, a disease or illness can be seen in quite a positive light, for it can point the person back towards the right path. Each person has a specific life plan or purpose and, once in tune with this and following it, the person's life flows much more easily and successfully.

We all have gut feelings, or what some people call an inner knowing, which try to help and guide us. If we choose to ignore this guidance and continue on the wrong path, these intuitions usually become louder and more noticeable until, finally, if we still fail to take notice, we may be, as the Americans say, "hit by a Mack truck", which, translated into Australian, means "run over by a semitrailer".

Yet look back at a time in your life when you have been flattened by a semitrailer, metaphorically speaking, and see if, in hindsight, there were not tremendous advantages that flowed on from that event. It probably represented a major turning point in your life.

The bush essences work by helping to keep your life running on course, by keeping you aware of and acting on your intuitions so that you don't need to be flattened too regularly by those semitrailers. And if you do experience a crisis, then the bush essences are available to help you work through your feelings and experiences and recover more quickly.

The bush essence remedies bring forth the positive qualities that reside deep inside every one of us. Their activation allows us to replace fear with courage, hatred with love, insecurity with self-confidence, etc.

Bach himself stated it quite wonderfully when he said that the function of flower essences is:

> To raise our vibrations and open up our channels for the reception of our spiritual self, to flood our natures with the particular virtues and to wash out from us the faults which were causing them. They are able, like beautiful music or any gloriously uplifting thing which gives us inspiration, to raise our very natures and bring us nearer to ourselves and by that very act to bring us peace and relieve our suffering. They cure not by attacking disease but by flooding our bodies with beautiful vibrations of our higher nature in the presence of which disease melts as snow in the sunshine.

Though the mechanism of their action has not been entirely proved, a great deal of research has been conducted into flower essences. Richard Gerber MD, in his excellent book *Vibrational Medicine*, reviews this research and presents his own theories. Gerber clearly shows that emotional and physical illness can be healed by balancing and treating our subtle-energy bodies—the astral, etheric, mental and causal bodies—along with our Higher Spiritual energies. These subtle-energy bodies play a major role in maintaining our good health.

According to Gurudas, the author of the book *Flower Essences*, of the three major forms of vibrational remedies—flower essences, homoeopathic remedies and gem elixirs—flower essences are the best and most effective modality to reach and treat the subtle-energy bodies, along with the meridians and physical body. He says that homoeopathic remedies generally operate at the physical level and on the biomagnetic fields of the body. Some of them can affect the chakras and subtle bodies, but not as effectively as flower essences. Gem elixirs act similarly to flower essences but not to the same degree, as they do not have the same potency of life force.

Gerber mentions that the term "vibration" is a synonym for frequency, and that the only difference between dense matter, such as an antibiotic or a piece of wood, and subtle matter, such as a flower essence, is the frequency at which they vibrate. Subtle matter vibrates at speeds exceeding the velocity of light. The vibrational medicines that contain high-frequency subtle energies are able to act on the subtle-energy bodies and at the level of the emotional, mental and spiritual vehicles.

Gurudas provides an interesting description of how the essences have their effect on the physical and subtle bodies. When an essence is ingested or absorbed through the skin, it is initially assimilated into the bloodstream. Then it settles midway between the circulatory and nervous systems. There, an electromagnetic current is created by the polarity of the two systems. The essence then moves directly to the meridians, which are vital mechanisms of interface between the subtle bodies and the physical body. From the meridians, the life force of the flower essence is amplified out to the chakras and various subtle bodies and then back again to the physical body. This amplification also magnifies the life force of the essence and aids in its assimilation. The essence reaches the imbalanced parts of the body faster and in a more stable form.

Flower essences, claims Gurudas, cleanse the aura and subtle bodies so that the imbalances will stop bringing about ill health. This cleansing occurs instantaneously, but the results take longer to show themselves. The quartz-like crystalline silica structures in the physical body, such as those in the bloodstream, the hair and nails, and in the subtle bodies amplify and transmit the healing energies of the flower essences to their appropriate sites of action, and at the correct frequencies. This whole process is similar to the way radio waves strike a crystal in a radio so that the crystal resonates with the high frequencies, absorbing them and transforming them into audio frequencies which can be heard by the human ear.

In addition, the following analogy is useful in explaining and understanding the effects of the flower essences on an individual.

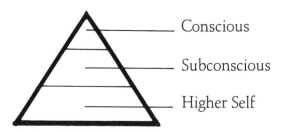

This diagram represents the psyche. At the top is the conscious mind, the part that thinks about what to have for lunch today and what you did last weekend, and that is analysing the very words you are now reading. The conscious mind is full of chatter, and it jumps constantly from one thought to another.

Beneath the conscious level resides the subconscious mind. There, many of our beliefs are stored. In fact, the majority of these were formed while we were in the womb and in the first few years of our lives. These beliefs often guide and direct our actions. We are constantly creating situations to reinforce the particular beliefs that we hold. We rarely have a conscious awareness of those beliefs formulated early in our lives or even later.

For example, at three years of age a child might have been told repeatedly by a parent or someone close to her that she is really stupid and unable to do anything properly. As she grew older she might have had no conscious memory of what was said to her but, nonetheless, that early message could have been stored as a belief. And that belief could arise frequently during her life. The pattern of her life would then be to create situations that reinforce the belief. For instance, if she got a challenging and responsible job, she would make silly mistakes and possibly lose the job. The frustrating part about the situation is that this person doesn't consciously know why she makes a mess of things and doesn't realise that her failures are caused by her belief that she is stupid and unable to do anything correctly. Observation would reveal that this pattern has occurred throughout her life, in fact, ever since she was three!

Other common beliefs are: that no one likes you; that you don't belong, especially if your parents did not plan or want to have you; that you are unattractive, which is often the result of influential people making offhand remarks or constantly belittling a child. A person's belief in their own unattractiveness could have been caused by an adult remarking to the child, "You ugly little blighter," as a rough form of endearment.

Obviously the intention of the person who speaks is relevant, but there is also tremendous power in the spoken word itself. Many people give themselves diarrhoea because they keep saying that things give them the shits. "You make me sick" and "I'm dying for a cigarette" both have a very literal meaning for the subconscious!

How many people do you know who, after their relationships break up, vow never to repeat that mistake again, yet immediately get involved with a more or less identical partner with whom they begin to re-enact the same scenario? The bevy of beliefs in their subconscious minds determines this pattern.

The loss of a parent can have a profound effect on a small child. A little girl who has lost her father in an accident may internalise this event into the belief that if you get close to men they will leave you. Later on, when she has relationships with men, they either lose interest in her, leave her for someone else, or head off overseas. She attracts men who will reinforce her belief by leaving her.

There are many ways in which negative beliefs can originate. However, it is now possible for people to liberate themselves from self-destructive behaviour patterns by releasing their negative beliefs. This will not only benefit them but will also set their children free from an often self-perpetuating cycle.

The above are just a few examples of negative beliefs, their causes and effects. Of course, there are also positive beliefs present in the subconscious.

Parents have a wonderful opportunity—and responsibility—to help develop these in their children.

Finally, the bottom section of the diagram represents the superconscious mind or Higher Self, whichever term you feel comfortable with. Here are the positive qualities inherent in us all—love, courage, trust and faith; the solutions to all our problems; and an understanding of our life plan and purpose. This part of the mind produces our gut feelings and our inspiration.

The Australian Bush Flower Essences work in a similar way to meditation in that they not only resolve negative beliefs, but also make those positive qualities within us come flooding through to our conscious minds. These powerful remedies work in a multitude of ways to help us let go of negative beliefs and emotions and replace these with the positive aspects of ourselves. People and events can suddenly crop up to help us move on to a new level of understanding. This may be further accentuated once we are no longer held back by our negative beliefs.

A person can be viewed as having a number of emotional layers, and the bush essences help to work through those layers of emotional blocks. The essences are self-adjusting, which means that they are effective only up to the point that people are ready to go. They also operate as gentle catalysts, taking people to that point and then helping them chip away at the next layer, too. Moreover, they are totally safe and without side effects.

Of course, the more self-awareness people develop through personal growth practices, meditation and a willingness to explore and confront their feelings, the faster the remedies will work.

The beauty of the essences is that anyone can use them as they are entirely safe and without side effects. You do not need years of medical training to be able to understand their properties and prescribe them. This book is presented in such a way that both the professional healer and those who want to use the remedies either on themselves or on their friends and families will be able to choose the most appropriate essences easily and quickly. The book serves as a stepping stone to the use of these profound remedies.

The bush essences are as effective when taken by children and animals as they are for adults. In fact, the essences work exceptionally quickly on children as they have built up less negativity and fewer emotional barriers and are more in touch with their feelings. The dosage for animals and children is the same as that for adults. The results achieved with young children and animals discount the possibility that the benefits of taking bush essences are due to the placebo effect.

The remedies are best taken singly in most cases, either as a single essence or in combination if the essences are all addressing the same theme, as then you will have a finer, longer-lasting effect. Then they will work from the outer bodies inward to the physical. When a combination of two or more is used, addressing separate issues, they may have a slower effect. However, in crisis situations, combinations of up to seven remedies can be used, though we normally suggest a maximum of four (or occasionally five).

Herbs, minerals, vitamins and cell salts work only on the physical body.

A
Bush Essence
—Journey—

In the spring of 1986 the heart of Australia saw its best wildflower display in decades. Because of the high rainfall in normally dry places like Alice Springs and Uluru (Ayers Rock), flowers that had not bloomed there for twenty years sprang up.

During 1987 my desire to explore the Centre become stronger and stronger. I had always had a passion for the bush, but now I felt a tremendous urge to go to the Northern Territory. So it was with great excitement that my wife, Kristin, and I planned our journey in search of Australian bush flower remedies, to be made up in the rays of the sun, as Dr Edward Bach did with his beloved English flower essences over fifty years ago.

As the departure date grew closer, I visualised the healing properties of three flowering plants that are found growing only in the Northern Territory. As with the majority of the other Australian Bush Flower Essences that I have developed, their healing properties were revealed to me while I was either meditating or in a quiet, reflective state.

On our first morning in Alice Springs we awoke staring up at the MacDonnell Ranges. Neither of us spoke for a long time as we soaked up the enormous power of this land. The magic of the journey had begun.

We spent the rest of that morning at the Olive Pink Flora Reserve. Olive Pink (1884–1975) was a remarkable woman, arriving in 1930 to live and study with the Aborigines of the Tanami Desert. Later, on moving to Alice Springs, she established the reserve with the aim of creating an arid zone botanic garden to display, conserve and research the flora of central Australia. Many rare and endangered plant species have been established here so they can be conserved and protected.

There are a number of gorges both west and east of Alice Springs. We chose to head west and spent our first night at Ormiston Gorge camped on the bank of the dry creek. Most of the creeks and rivers in the centre are now dry sand beds dotted with ghost gums. One good fall of rain in spring can almost immediately bring out a sea of wildflowers. That night we heard a storm build up, and then down poured the rain. We were ecstatic—our first night there and the drought had broken.

As only 4 millimetres fell that night, the creek was still dry and there did not appear to be any more flowers around. Our long faces quickly brightened at the sight of Tall Yellow Top (*Senecio magnificus*), which was flowering all around us, and the perfect weather—a cloudless blue sky, the perfect conditions for making up essences.

Becoming sensitive to the plant's healing properties that morning was simple. *Senecio magnificus* is for alienation, for an individual's loss of connection to his family, peers or country. There is a sense of "not belonging" with this remedy. Often the head has taken over from the heart, to compensate for the alienation, and there is a loss of the heart connection. This is a very special Australian Bush Flower Essence, and one I feel will be commonly used.

The next part of our trip was spent travelling further west, exploring the many gorges and then finally Palm Valley. The country we saw was truly amazing. We often had to stop the car to take in the beauty, strength and timelessness of the land. Yet those few days were also very frustrating.

Feeling the energies of the land and finding so many flowers, I started to worry that we were only there for a limited time and I had better make up as many remedies as quickly as I could, as who knew when I would be there again. A sense of urgency and panic came over me. I then decided to let go of my impatient mind that was trying to control events and instead to have trust and let things flow—as they invariably do.

That was the turning point of the trip for me. As my need to rush and my feeling of insecurity left, I was able to experience whatever happened without any worries at all.

My four-wheel drive apprenticeship was completed during our days in Palm Valley. The 16 kilometres of dry, deep, sandy river bed of the Finke—the oldest river in the world—was just an entree to navigating the jagged-rock road leading into Palm Valley. The valley is an oasis in the middle of this arid land. Towering walls dotted with cycad palms and ferns encased the remains of Australia's long sought-after inland "sea". Unfortunately for the tragic explorer Ludwig Leichhardt, it had existed millions of years earlier. Some of the world's rarest plants, including the cabbage palm, are found in this valley. The cabbage palm roots tap deep into the artesian bores trapped between the rock strata.

Our first sight of the Sturt Desert Rose occurred on a dirt track leading to Palm Valley. I particularly wanted to make up this flower essence. The beauty of the flower confirmed my belief that it would be a significant remedy. Unfortunately this shrub was too close to the track to use. We would have to wait a little longer. In fact, it was not an easy flower to find untarnished.

The head ranger at Uluru, our next stop, gave us helpful information on where to find the Sturt Desert Rose and allowed us to search in areas that had been put off limits to allow the land to regenerate. One and a half days later we still hadn't found it. In fact, another ranger told us of the area where we eventually found the Rose. However, those thirty-six hours were not wasted. Every moment I spent in the Olgas will always be treasured. Of all the special places in the Territory, none touched me quite as much as the Olgas.

I felt such satisfaction standing in front of the Desert Rose as the pink flower gently waved in the breeze, but I felt awe too. Here it finally was, after all that searching. In a very short time there would be an alchemic reaction with the Desert Rose, producing a healing essence which people would experience for years and years and which would be a catalyst for their development and health.

Through talking to the rangers—both white and Aboriginal—Kristin and I had been learning of the Aboriginal *Tjukurpa*, a term that had been misconstrued by early white people as meaning "dreaming" or "dreamtime"—a totally inadequte translation.

The Anangu Aborigines are the traditional land owners of this area. To them, their law, their philosophy and their connection to the land is through the *Tjukurpa*. The *Tjukurpa* is existence itself, in the past, present and future, and also the explanation of existence. It is the law that governs all behaviour. The *Tjukurpa* is still being recreated and celebrated by the Anangu today. The Anangu say their country was made during the *Tjukurpa* and it was then that places such as Uluru (Ayers Rock) and Katajuta (the Olgas) were named.

Uluru has recently been given back to the traditional land owners who lease it back to the government for a comparatively small sum. It is their land, which they have agreed to give non-Aboriginal people the chance to see.

As I meditated on the Desert Rose I knew I couldn't proceed with making it up unless I had permission from the Anangu. A short time earlier a ranger had driven up and stopped. I had met that particular ranger the day before and there was a good connection. My gut feeling then was to ask him if it would be all right to pick a couple of the flowers; but after all the efforts to find the Desert Rose, what if this was the last one still in flower. I held back, as I did not want to take the chance of him saying no. Well, it was too late now; the ranger was gone. How quickly the emotions and intellect can negate one's gut feeling.

The result was an 80-kilometre round trip on a corrugated red-dirt road to the Ranger Station at Ayers Rock and back again. There was no question of not going; we had to go, otherwise the energy in the essence would have been incomplete—it would not have worked. Our problem was time. We determined how many minutes of sunlight the Desert Rose needed to be ready. By our reckoning, taking into account the time of the sunset, we had at the most forty-five minutes in which to make the round trip and still have enough time left to attune ourselves to the Rose and make the essence up. That's 110 kilometres per hour on a dirt road.

"Let's do it," yelled Kristin, as we hurled ourselves into the Landcruiser and hooned off. John Belushi and Dan Ackroyd, *The Blues Brothers*, had nothing on us as we sped, jolted, bounced and flew over the corrugated road with red dust ballooning up behind us. The ranger said "yes".

Time and again on our trip as we made up a remedy we would become enmeshed in the property of that flower. Sturt Desert Rose is for guilt, for not following your own inner convictions and morality. It helps you to choose what you know you have to do; it can bring back self-esteem which has been destroyed by actions in the past that you feel great guilt about.

The sunset that night blared across the heavens for what seemed like hours, bathing the sky, clouds and the Olgas in sublime, breathtaking colours and patterns. It was also the night after the full moon, and a huge orange moon broke over the horizon that evening. That night and the night of the full moon at Ayers Rock we had stayed out and had very long and deep meditations. On both evenings time seemed to disappear. We decided to celebrate the making of the Desert Rose and splash out at the Yulara Sheraton Hotel for a late dinner. As we dropped our dusty bodies into the elegant surroundings, we noticed the name of the restaurant—The Desert Rose.

Our old friend, the head ranger at Ayers Rock, assured us that we would find a 4-kilometre stretch of road past Kulgera that was lined with flowering Sturt Desert Pea. Seventy kilometres after passing Kulgera we found ourselves in South Australia with no Desert Pea to report!

The locals back at Kulgera weren't too sure where we would find this plant, but some mentioned that it had been seen a couple of years before, 25 kilometres along the railway line. Apart from that we would have to travel into the Simpson Desert, which seemed too far away and too rugged.

Instead, we decided to follow our gut feelings and head up one of the dirt roads leading out of Kulgera. It took us past abandoned cattle yards and the Kulgera railway station—also apparently abandoned. A little further on were eight houses, all in the same government style, dropped starkly in the middle of kilometres of red dust. Here we eventually tracked down a very warm and friendly man from Yugoslavia whose work took him along the railway line, where Sturt Desert Pea was definitely growing. One of his friends thought we were crazy heading off just before sunset as it would take us nearly three hours' travelling along a terrible track to get there—and all for a few flowers.

Inspired by our previous Blues Brothers-style "mission from God" dash in the Olgas, we were already roaring down the track after our new grail. The track follows the Ghan railway line through to Alice Springs. This land is truly in the middle of nowhere and at times totally barren and flat for a full 360-degree view. We raced with the setting sun to get there before dark. At the point where the sun dipped behind the railway line right on sunset we came to our landmark, and there, far off in a ditch, was the Sturt Desert Pea, just as we were told it would be. We pulled out our chairs and meditated around the powerful flowers, alone in the middle of the desert, just us, the Desert Pea—and the flies.

Later that night as I drove on, I realised that our fuel was low; there was not enough to turn back to Kulgera and maybe not enough to make it to Alice. Most of the territory is unfenced and driving during the night is dangerous. Just keeping on the track was difficult enough, let alone watching out for animals.

Suddenly the track turned left, under the railway line and into nowhere. After a few minutes the road literally dropped off. There was no more track and we had no idea where we were. Near the railway line we had passed a generator and a light, so we headed back to find the light coming from outside a windowless caravan. I cautiously got out to check it out but Kristin was quite worried.

My knock was answered by a Slavic man in shorts, who opened the door to reveal thousands of moths hovering around the lights inside his caravan. He was not in the least perturbed to find someone knocking on his door in the middle of the night and helped us find our way. At this point Kristin was feeling scared by the unknown and the dark and I was worried about the low fuel gauge.

We turned on the radio for the first time in days, to ease the tension, only to hear the end of a beautiful piece of classical music—a Bach chorale! Then the airwaves were filled by the soothing tones of Caroline Jones. She was interviewing Brother Andrew who, with Mother Teresa, had started an order of Brothers to help the poor in India. He is a very gentle, humble, spiritual man who speaks beautifully. He claimed that the materially poor in India and Asia are often happy people and that he enjoys working with them. What deeply upsets him is the suppressed anger, pain and sadness in the average Western affluent person and the fact that there is a spiritual desert in so many of these people. There we were driving through the desert, searching for remedies to address this spiritual poverty, and on hearing his words we were overcome by the incredible power and beauty of that moment!

The other guest Caroline Jones interviewed was a woman who had survived a horrendous childhood in Europe, including the Second World War. She had always been fascinated by Australia and the Aborigines. She related to their alienation and adopted their culture and struggle in Australia. As she discussed the beauty of the outback, we were there, driving through it. She spoke of the spiritual significance of the moon to Aboriginals, and at that very moment the moon rose on the horizon! Kristin had tears streaming down her face and we both felt that this whole experience was one of the most powerful in our lives.

As you may expect, the power of the Sturt Desert Pea matches the energy in which its essence was conceived. It is the remedy for removing very deep hurts and pains.

On our long journey north, driving stints were broken to photograph numerous flowers and to explore interesting dirt tracks. We spent one night at Wauchope mesmerised by the beauty and sheer number of shooting stars.

The landscape slowly changed before us. The short, stunted growth of the Centre and the yellow carpet of *Senecio magnificus* changed to taller plants and trees, blown into twisted shapes by the year-round hot wind.

Between Alice Springs and Katherine we made up Wild Potato Bush which is for slow and heavy people who are ready to move on. These people feel as if they just want to unbutton the old self and step out. They find moving forward very difficult, and this remedy will help them do it.

Spinifex grass—my first grass—was developed slightly differently from the rest. It can be used topically for blistering skin conditions such as herpes and for clean cuts made, for example, during surgery or by glass. Taken internally, it brings to the surface the emotional states and beliefs that are creating the herpes or skin conditions. There are so many people who could benefit from such a remedy.

The nights were quite warm by the time we reached Katherine Gorge, which in fact consists of thirteen amazing gorges linked together. We swam and boated up to the seventh gorge, washing off as much accumulated dust as possible and doing as much swimming as we could while still in safe waters away from the salt-water crocodiles.

It was in Katherine that we finally came across a flowering *Calytrix exstipulata*, or Turkey Bush. We had been hoping for this opportunity. The plant flowers in July and August, and we knew before leaving Sydney that the flower is for creativity and for helping to clear emotional blocks and discouragement, which hold many people back from creativity when they lose inspiration. It helps to remove distraction and frustration and allows them to channel creativity.

The final leg of the journey was Kakadu National Park. So many people we'd met on the road told us of their disappointment with Kakadu, describing it as hot, dry and greatly overrated. For us, this tempered

Australian artist Charles Blackman's description of it as the Eighth Wonder of the World.

We entered Kakadu via Pine Creek in the early afternoon. As we got out of the four-wheel drive, we met with a furnace of heat. We quickly opted to head down to UDP Falls and relief by the water. We had decided in Katherine not to swim in Kakadu, as one is never totally safe from the salt-water crocodiles. Down at the falls, however, there were scores of school kids, most of them splashing in the water. Initially being careful to stay in the midst of the children for safety, we soon let go of our resolve and cooled off. That night we went spotlighting crocs along the South Alligator River. You can see two bright orange points of light staring at you from the other bank when the beams catch their eyes. We counted fifteen that night, all of them believed to be the smaller Freshwater Johnson Crocodile—an animal afraid of people.

At the northern end of the park is Ubirr (Obiri Rock) where the magnificent Aboriginal rock paintings—the oldest in the world with some dating back 23,000 years—are to be found. This major art site is the only one known in the world to show various painting styles over such a long period of time. The ranger who took us there told us to imagine we were entering a great national gallery—she was certainly right.

From the top of Obiri Rock you get the best view of Kakadu. There we saw a sunset that for both of us has never been surpassed, a common claim from those who have been to the rock.

Every morning we would wake early, filled with excitement in anticipation of the new day. The morning after we visited Obiri we found Billy Goat Plum growing next to the tent. The remedy made from the flowers is for self-disgust and dislike of the body, especially the sex organs. The people needing it often feel dirty after sex.

As we headed back to Obiri Rock, the scent of gardenia wafted to us through the still heat. We then spotted the source, Bush Gardenia, not the cultivated suburban shrub but a 9-metre tree! There were three trees side by side, in full flower. The flowers only last a day or so in the heat of the late dry season. What a morning for essences!

As Kristin painted the scene I prepared the Bush Gardenia essence. This remedy is for fading passion. It helps to recreate the passion in relationships, but, more important, it restores a person's interest in their partner. The essence can also be used by a family in which one member is going off the rails while the others are too caught up in their own affairs to take much notice.

The final remedy made during the trip was the Red Lily. We spent the day we made it up in the company of Mick Alderson and a mate of his from Darwin, Don. Mick is a traditional land owner of Kakadu and has worked as a ranger for twenty years. Mick took us to the wetlands in the centre of Kakadu, which is the breeding ground for millions of birds. Australia has over 670 species of birds and 270 of them are found in Kakadu. Mick and Don seemed to know the names of all of them as they pointed them out to us.

Mick took us to a billabong which he knew would be full of Red Lilies. It was reassuring to be with this man, who was so familiar with the land. As we waded through the billabong to the lilies, knee-deep in mud, his words, guaranteeing that there were no crocodiles and that the mud would not rise much higher than our knees, calmed us. Half-way to the lilies, Kristin found leeches all over her feet and quickly lost her enthusiasm for the second half. This lily, also known as the Sacred Lotus, makes an excellent vegetable. The fruit, stems, leaves and flowers are all eaten. The seeds are very high in fat and protein and keep well. The rhizomes have a high fibre content, are a good energy source and taste very sweet. The stem can also be used as a drinking straw, to tap the clean water under a layer of scum and to clean any dirty water as it passes along the length of the stem. As we waded out to the lily we disturbed hundreds of Magpie Geese, which rose in spectacular flight.

While waiting for the Red Lily essence to be ready, we watched a jabiru, Australia's only stork, catch a 1-metre black snake and eat him head first.

"How many snakes have we walked over in the mud?" we asked. Mick just laughed.

Kakadu was certainly a very fitting finale to such a magical trip. We felt the God force flowing strongly through us and the land as we travelled through the Northern Territory. I highly recommend that you explore and experience this part of Australia.

The Magic
of
Making Bush
—Essences—

The purpose of this section is to share some of the remarkable events that have commonly occurred while making up the Australian Bush Flower Essences. I have chosen just a few of the different remedies to highlight the undeniable presence of guidance which was so evident during the making up of all the remedies.

We chose the Waratah as our logo because of the tremendous importance of its properties, especially in the years to come. Yet in the first year of making bush essences, the Waratah's flowering season was coming to an end, and we still hadn't made an essence from the plant.

The coastal variety had finished flowering weeks earlier, so we were relying on making the remedy from flowers on the other side of the Great Dividing Range, where the plant comes into season much later. I had been in regular contact with a man who collected Waratahs commercially on his property. Every two weeks or so I would phone and ask him if they were now in flower, and he would give the same reply: "No, the frost has been bad this year, and they are taking longer than normal. Give me another call in a couple of weeks and I will let you know if they are ready then."

After a while I became somewhat frustrated by the lateness of the Waratahs that year and just stopped ringing him, believing that I still had plenty of time before they bloomed. It came as quite a shock when I finally called him and heard that they had already flowered and were now finished. He doubted that I would find any left in bloom, but he gave me the phone number of another grower, from whom I received the same grim news along with yet another phone number.

This whole process was repeated a few more times until, finally, via John Dixon, a university researcher and author of a definitive book on Waratahs, I was given the phone number of a company which reputedly knew the whereabouts of the last flowering Waratahs in the State. They agreed to tell me where this special spot was on the strength that I had come through John Dixon and only after they had called a special company meeting to discuss whether or not they should divulge this legendary location to me. They were concerned because their whole livelihood would be at risk if other people were to plunder this spot. They informed me that I had permission to go there, though they doubted very much whether I would find any Waratahs in flower. They had been to that area a week earlier and had picked the last remaining flowers.

The next day was the only opportunity I would have to make the four-hour journey, and that night, as I went into meditation, I simply asked for guidance and direction in my search for flowering Waratah. What followed filled me with great hope for the trip ahead, because I was shown, in meditation, a steep gully leading down to a creek bed where there was a flowering Waratah.

Lo and behold, when I arrived at the location the next day it looked exactly as I had seen it the night before. The gully itself was full of Waratahs—over 300 of them. From afar they all looked as though they were in perfect bloom, but on closer examination I saw that they had started to wither and fade, having been slightly burnt by the sun.

An aura of power and majesty pervaded the whole scene, and I took my time wandering among them, taking in the magic of that wondrous place, till eventually I came upon the very same spot I had been shown in meditation—and there was my Waratah in full, glorious bloom.

After letting out a few exuberant hoots of delight and giving thanks to all my unseen helpers, I prepared the essence. The weather wasn't ideal for flower-essence-making—there were clouds in the sky and rain was hovering about. Yet an area 3 metres in diameter around the bowl containing the Waratah was bathed in full sunlight for the whole time it was needed to make up the essence. And at the very moment I took the Waratah out and decanted the essence, the heavens opened up and rain poured down. It had held off just long enough to allow me to complete the task.

I took shelter, waiting a few hours for the rain to stop, and later, as daylight began to fade, I was drawn back for a final wander among those proud plants. Then I came across another flowering Waratah which I also picked and made up into an essence. Later that night, while celebrating the wonderful events of the day, I realised that it had been picked at the precise moment of the full moon, and that the essence of the second Waratah was made up in moonlight only.

The making up of Paw Paw involved a similar story. In the tropics on a day with the temperature in the high thirties, I was preparing the essence in a field into which a number of cows often wandered. As I was in the middle of a workshop, I had to abandon my desire to stay with the essence, though I was more than a little concerned that the cows might feel that all their Christmases had come at once if they were to discover a bowl of cool water with luscious Paw Paw flowers floating in it.

My only recourse, as I saw it, was to ask for protection to be placed around the bowl by Spirit, to visualise white light surrounding it, and to trust. On returning hours later, I found a number of cow pats close to the bowl, but for some reason or other—well, I think I know why— the cows hadn't actually disturbed or drunk the water or eaten the flowers—something had stopped them. It was again with thanks that I decanted the essence.

Bush Iris was the first remedy that we ever made. We had originally set out that day with the intention of making up the Waratah essence but the plants weren't in perfect bloom. As there were scores of Bush Irises in flower and we were aware of their properties, we chose to make that essence. When we returned to the same spot the next day, we were astonished to find absolutely none of them left in flower—a most unusual occurrence. Both Kristin and I had a strong feeling that the Bush Irises had been waiting just long enough for us to come and distil their healing essence for humanity.

As for Silver Princess, we were told that only one naturally occurring stand was left in the whole of Australia. And though it wasn't the usual time for these trees to flower when we made the journey to the lonely spot at Boyagin Rock in the south-west of Western Australia, we found not only the stand but also its beautiful trees in flower. As we wouldn't have had another opportunity to go to this area for a few years, we had ignored the advice of many botanic experts that we were wasting our time. I never once doubted my strong inner knowing that we would find the flowers and make up the essence of Silver Princess.

It was extraordinary how the Red Helmet Orchid came into being as an essence. I had already received the sense that there was an orchid that had certain properties, but it was a matter of being able to find it. We had been searching for "the Orchid" with those properties for quite some time when finally the Universe pointed it out to us in the Stirling Range. This was achieved by a car pulling up opposite our vehicle on a small dirt road leading up to a mountain walk. A fellow jumped out and ran across to me in the bush, where I was photographing, to tell me excitedly that he had just found a flowering Red Helmet Orchid half-way up the trail to Toolbrunup Peak. He said that he had to come and tell me because he knew that I'd be just as excited as he was! This man took the time to stop, get out of his car and race 50 metres through the bush to tell me about the orchid, as if he knew that he had some very important information for me. He was certainly a wonderful Mercury and bringer of good tidings.

At that point I realised that this was the orchid we were looking for, and I felt absolutely certain about this. It is an exhilarating experience to tap into such certainty, and to know that we can be guided by the unseen and benevolent forces all around us.

Using Bush Essences with Other —Modalities—

Numerology

Numerology is a very old system of analysis which provides a simple method for quickly and accurately gaining an understanding of an individual's personality and abilities. It reveals a person's strengths as well as identifying areas of potential difficulty. Numerology is a very useful tool for helping to select the appropriate Australian Bush Flower Essences, even indicating some that might not have been obvious.

There are many different systems of numerology. The one I will be describing here is the Pythagorean system which is based on the birthdate and was devised more than 2500 years ago. The analysis of any birthdate will indicate the particular remedies that are potentially appropriate.

The first step in this system of numerology is to write down the birthdate. Each number in the birthdate has its specific place on a grid and is always stored there. The grid appears as follows:

3	6	9
2	5	8
1	4	7

For example, if we were working with the birthdate 9 November 1942, we would express this as 9.11.1942. Then each of the numerals would be written down in its specific place on the grid:

		99
2		
111	4	

The top line of the grid (3—6—9) represents the mental plane, the middle line (2—5—8) the emotional plane and the bottom line (1—4—7) the physical plane. Each place on the grid represents a specific characteristic or aspect of personality. The total number of numerals in a specific place on the grid determines the degree or intensity to which that characteristic may be expressed.

The Number 1

A single 1 on the grid, or birth chart, indicates a potential difficulty in communicating one's feelings to others. These people often assume that others know what they are feeling without being told. Of course, other people usually don't know what these people are feeling, so misunderstandings often arise in their relationships with others.

Most people with a single 1 feel very self-conscious when talking about themselves as they believe that they must be boring their listeners or that they are being self-centred. In fact, a closer bond may develop between two people when they trust each other enough to reveal their feelings.

The Philotheca remedy will help these people speak about themselves, and Flannel Flower will help them share their innermost feelings.

Two 1s indicate the potential for the easy expression of one's feelings.

Three 1s indicate either a very chatty person or a very shy, quiet person.

Kangaroo Paw is the remedy for very talkative people as it will help to make them aware of the needs of others.

The very quiet people hold back with others, rarely opening up until they feel more comfortable and safe with them. Five Corners will help them develop self-esteem and Dog Rose will help them feel more secure with others.

A person with four 1s may have great difficulty in expressing emotions and in communicating with other people. They often require quiet times alone and are often misunderstood by others. Two of the remedies that relate to this pattern are Flannel Flower and Bluebell for the expression of feelings.

Someone with five 1s has even more difficulty in communicating. This person will benefit from taking the same remedies as a person with four 1s, but may also need Bush Fuchsia for clarity of speech and voicing their opinions.

The Number 2

This number represents sensitivity. A promising thought is that in the year 2000 and for the next thousand years every person born will have a 2 in their birth charts. This fact indicates that greater sensitivity and intuition will be shown by future generations.

People who have two or more 2s in their birth charts have a very high degree of sensitivity which may cause them to be very easily hurt. They may be wounded by actions and comments that others would ignore. Remedies that can help them are Sturt Desert Pea for long-held, deep sadness and unresolved grief; Crowea to encourage a calm and centred state; and Flame Tree to ease feelings of rejection.

The greater sensitivity of these people may be manifested as heightened empathy. As a result, they may act as psychic sponges, absorbing the feelings of others. Yet they will not be aware that the feelings that they have aren't their own. I commonly find that these people are surrogates. They display the symptoms of other people's illnesses. For example, if they massage someone with a sore shoulder, they will also develop a sore shoulder. An excellent remedy for closing themselves off and protecting themselves from other people's thoughts and feelings is Fringed Violet.

People without a 2 in their charts are not necessarily insensitive. Everyone can develop a particular quality in themselves, even if they weren't born with it, though it is more difficult to become a sensitive person if you don't have a 2 in your birth chart. The numbers represent inherent potential only, and by using the bush essences and practising other forms of personal growth, the desired qualities can be developed. They also pinpoint latent skills and abilities that are not being used, so that once people are aware of these, they can choose whether or not they will develop their potential.

The Number 3

This is the first number on the mental plane and deals with imagination and mental focus. People with no 3 in their birthdates would benefit from some form of mental discipline which exercises their minds, including the memory, as they tend to forget detail. Appropriate remedies are Isopogon for learning and memory; Bush Fuchsia for balancing the left and right hemispheres of the brain; Sundew for attention to detail; and Cowslip Orchid for detached judgement and objective analysis.

If the imagination is not very well developed, then a remedy such as Little Flannel Flower will bring a sense of playfulness and lightness to the person. Turkey Bush for inspiration and creativity may also be appropriate.

The more 3s there are in a chart, the greater the imagination. Many science fiction writers have a number of 3s. However, if a chart has more than one 3 but there are no numbers next to the 3s, that is, no 2, 5 and 6, then the person may have a slight paranoid tendency stemming from a vivid imagination. In which case Mountain Devil for suspicion is quite an appropriate remedy. Sundew can also be used by those who are constantly lost in their imaginations—the daydreamers.

The Number 4

This number is at the centre of the physical plane. People with a number of 4s can be a little too materialistic, in which case Bush Iris will help. Sometimes they show an accompanying insensitivity to the needs of others. The remedy for this is Kangaroo Paw, or Red Helmet Orchid if they are neglecting their own children in their pursuit of wealth or power.

Many 4s can also indicate a great deal of practicality, though care should be taken to balance this by using the intuitive and cognitive abilities. Bush Fuchsia for developing intuition is a good remedy for this configuration.

An abundance of 4s can also give rise to a harsh outlook, as a result of seeing life from only a practical or functional point of view. Turkey Bush can open up a person to the beauty of creation and nature. Hibbertia can be taken if the person is too rigid and strict with themselves. Any inclination to excessive tidiness, which is common with a number of 4s, can be balanced by Boronia.

No 4s can indicate someone who gives up easily, especially if there are also no 5s or 6s in the birthdate. Kapok is the remedy here, as it encourages perseverance.

No 4s in a chart which has 1, 5 and 9 shows great determination, though the person can easily become very impatient. The remedy for this is Black-eyed Susan.

The Number 5

This number is in the centre of the emotional plane. People with no 5s in their charts are apt to be a little off-centre with their emotions. They are not totally sure what they are feeling so tend to react inappropriately to situations. Crowea is a very good remedy for this.

If there is more than one 5, especially three or more, the person invariably has great emotional intensity. You can often feel these people even before they come into a room—Black-eyed Susan. Their intensity usually results in their being self-centred as well as very full on, so Boronia (obsessional) or Kangaroo Paw (insensitive or gauche) may be appropriate. If their moods go up and down, then Peach-flowered Tea-tree will help. Rather than using alcohol, drugs and sex as a release for their pent-up intensity, they could try Bush Iris instead.

The Number 6

The number 6 deals, on one hand, with creativity and, on the other, with procreativity and sexuality. People with more than one 6 are often extremely aware of and very sensitive to their environment. If there is any disharmony, they will automatically feel "out of whack" and ill-at-ease. Especially when there is tension between their parents, these children can become quite ill, usually with stomach-aches, headaches and bronchitis. Crowea will address worry about the home and family. If there are three 6s, these worries can become obsessional—Boronia.

People with one or more 6s generally have a great need for rest which, if not met, may require a dose or two of either *Banksia robur* or Macrocarpa to restore depleted energy and vitality.

For people without 6s in their charts, Turkey Bush is a wonderful essence for developing creativity.

For a full description of the many bush essences that are appropriate to sexuality, see the chapter entitled "Sexuality".

The Number 7

This number represents learning and sacrifice. Many lessons are learnt by people who have a number of 7s in their charts. They often experience what appears to be much sacrifice and hardship in their lives with money, health and relationships frequently lost. These are lessons to teach a person detachment from material possessions and to bring about the realisation that there are more important things in life, an understanding that is also provided by Bush Iris. It seems as if people with a number of 7s in their charts choose to have quite a lot of learning experiences of the semitrailer variety in their lives. The Waratah, Fringed Violet and Paw Paw remedies are all excellent for coping with the arrival of these lessons!

These individuals don't like being taught by other people, preferring to experience lessons first-hand. If they are very headstrong and don't listen to the advice of others, or perhaps demand that everyone does what they say, Isopogon is appropriate. Isopogon can also be taken if they keep making the same mistakes without learning from them. Sunshine Wattle can be used to provide a glimmer of light at the end of the tunnel for those with many 7s in their charts, who may be feeling that life is just one big struggle. Or Southern Cross may be taken if they feel resentful that they've been unfairly singled out for all these lessons one after the other.

Quite often major events occur in their lives (Bottlebrush for going through life's transitions) at the ages of seven, fourteen and twenty-one.

The Number 8

This number represents organisation and control. People with at least one 8 in their charts can often be controlling—Isopogon. Two or three 8s can indicate that these people are also very restless. They can't seem to find their life path or plan, or don't even realise that they have one. Silver Princess can be of assistance to these people. Or Jacaranda, for those who are always rushing around, frequently changing jobs, abodes, friends and interests. For any lack of follow-through or enthusiasm typical of a number of 8s, Peach-flowered Tea-tree is helpful, while Wedding Bush will enable them to commit themselves to projects and other people.

The Number 9

Common to the charts of everyone born this century, 9 is the number of humanitarian concern and responsibility. These people can be very idealistic, more so if there are a number of 9s in their charts. Furthermore, if there are no other numbers next to these 9s, that is, 6, 5 and 8, the person is not only very idealistic, but also probably very impractical, very pie-in-the-sky. In this case Red Lily or Sundew may be appropriate, or even Kangaroo Paw if that impractical idealism translates into naivety. Hibbertia, too, could be thought of if the person was fanatical.

The number 9 also represents ambition, and if those people with a number of 9s are negative, they can be very ruthless in their ambition. Isopogon can be used for a controlling, dominating nature, or Bluebell for those not willing to share and wanting to grab everything for themselves.

Those with two or more 9s must be careful not to be too judgmental or critical, and to foster the ability to view situations and people either objectively—Yellow Cowslip Orchid—or intuitively, for that matter—Bush Fuchsia.

Configurations

The combination of three numbers in a straight line, whether 1—5—9 diagonally, 3—6—9 horizontally or 1—2—3 vertically, indicates a potential quality in addition to the qualities of the three individual numbers. Most charts have at least one of these arrows.

An interesting configuration is the arrow of hyperactivity, where the numbers 7—8—9 appear in the birth chart. With this arrow, children, in particular, need a lot of love and can be very trying for their parents as they can easily become a bit manic. They pick up on emotions quickly and feel things very intensely. Black-eyed Susan can be a good remedy for these children, though many benefit more from Jacaranda because their concentration and attention span is so short. They require heaps of TLC, especially from their parents and other adults in their life.

The numbers 4—5—6, when all present in a chart, are called the arrow of will. Usually it is the head rather than the intuition ruling the will, so remedies such as Hibbertia, Tall Yellow Top and Bush Fuchsia, which all work to balance the head—or the intellect—with the intuition, are often dispensed.

If the person has 1—5—9 as well as 4—5—6 there can be great intensity of purpose, especially when the life direction has been found, or a cause has been adopted. These people can pursue their goal or cause with great intensity—Black-eyed Susan—to the point of becoming unaware of who or what they might be trampling over in the process—Isopogon and Kangaroo Paw. Until they find their path they will experience immense frustration—Silver Princess.

The omission of any three numbers in a straight line also indicates certain potentials. No 3, 5 or 7 is referred to as the arrow of scepticism, and these people, at certain times especially early in their lives, are likely to be very resistant or sceptical to things of a metaphysical nature. Usually they readily embrace the orthodox, conventional stand on questions involving science and religion. Bauhinia may be appropriate for becoming open to and accepting new ideas and opinions. On the positive side, they are unlikely to be gullible or taken for a ride. Perhaps Bush Fuchsia for the development of intuition and Bush Iris for faith and the realisation that there is more than the mere physical—or what the five senses can perceive—could prove useful. These remedies could also be used by people with the 3—5—7 present, the arrow of spirituality, if they are not fully utilising all their skills and abilities in these areas.

The arrow of frustration prevails when there is no 6, 5 or 4 in a chart. I've invariably found that these people have suffered from tremendous hardships and sadness in their lives, perhaps business projects failing, health problems, relationships not working, losing family members or their homes, and certain remedies seem to offer a great deal of solace and acceptance at these times—Tall Yellow Top and Sturt Desert Pea. Wild Potato Bush and Spinifex can be taken if the frustration arises from physical incapacity caused by paralysis, MS, quadraplegia or an "incurable" disease such as herpes.

If these people feel resentment or believe they are victims of circumstances, then Southern Cross will move them closer to the understanding that they are the creators of those events, and that those situations may be seen as desirable if these people would consider the advantages that they offer and that they have already brought to their lives.

Some people, in response to unhappiness, blame others or bear grudges— Mountain Devil. Perhaps a relationship has broken up and resentment is felt towards the person who ended it—Dagger Hakea, whereas Bush Gardenia may improve the quality of a relationship so that neither party wants to end it.

Finally, when there is no 2, 5 or 8 in a chart, we find the arrow of hypersensitivity, where people are very easily hurt. These children constantly wear their hearts upon their sleeves. Unfortunately, many learn early to cope with the pain by cutting off. They build a shell around their soft, gentle, sensitive selves as a protection. In the process, they also lock out those who would care for them, and block off their own feelings. These attitudes and patterns can be released with Bluebell, for opening the heart, Red Grevillea, for overcoming sensitivity to criticism, and Illawarra Flame Tree for overcoming feelings of rejection.

Personal Year Cycles

Other insights can be derived from an understanding of the Personal Year Cycles. Each person has a nine-year cycle in which the energy, or potential focus, changes from year to year. By working with this cycle, as many people intuitively are, one is able to develop specific skills to a higher degree in a particular year than in any other year in that nine-year cycle. For example, if a person during a Personal Year 6, which focuses on creativity, chose to study a musical instrument, they would learn to play much more quickly in that year than in any of the other eight years of the cycle.

To determine your Personal Year number in relation to this nine-year cycle, simply add the day and the month of your birthdate to the current year. My daughter Grace's birthday is 28 November, so if I wanted to work out her Personal Year number for 1990, I would add 2+8+1+1+1+9+9+0 = 31. Because we need a number between 1 and 9, we further reduce any two-digit number to a single digit. This is done by adding the two digits together, so that 31 becomes 3+1 = 4. Grace is in a Personal Year 4 from January 1990 until 31 December 1990. Note that not everybody will be in a Personal Year 4 in 1990. Someone born on 17 March will be in a Personal Year 3 in 1990. That is, 1+7+3+1+9+9+0 = 30; then 3+0 = 3.

The major theme of each Personal Year number from 1 to 9, as well as some of the bush essences that help to maximise the potential of each Personal Year number are as follows:

Year 9: A year of changes, whether in relationships, career or home, which usually work out well.
Bauhinia, Bottlebrush, Red Grevillea, Silver Princess.

Year 1: The changes in Year 9 often bring rewards this year. This is also a good year to make changes.
Same remedies as in Year 9.

Year 2: Working in partnership or in cooperation with others; developing sensitivity and intuition.
Bush Fuchsia, Bush Iris, Slender Rice Flower.

Year 3: A good year for mental stimulation—study or travel, especially to a different culture.
Bauhinia, Bush Fuchsia, Isopogon, Paw Paw.

Year 4: A year to take stock of and consolidate all the changes and events of the last four years; and to nurture and recharge the body, especially the nervous system. Anyone who continues to push themselves too hard in Year 4 is putting their health at risk.
Banksia robur, Black-eyed Susan, Bush Gardenia, Jacaranda.

Year 5: A great year for working through any backlog of emotional rubbish. Personal growth courses, etc, highly recommended.
Bottlebrush, Crowea, Flannel Flower, Hibbertia, Bluebell.

Year 6: Clearing out all the emotional rubbish in Year 5 has opened the way for an important relationship. Many couples either meet or commit themselves to one another in Year 6. Established relationships can also get a boost. Very good friends are made in this year. More energy is available for creative pursuits.
Bluebell, Bush Gardenia, Turkey Bush, Wedding Bush.

Year 7: The year of the semitrailer!—and for learning major lessons. What you have that isn't really yours can be taken away. Relationships, health and wealth, to name only a few, receive the acid test in Year 7.
Fringed Violet, Kapok Bush, Southern Cross, Waratah, Isopogon.

Year 8: The tools needed for making changes in Year 9 come to you. Generally, this is a very good year financially. You can more easily organise yourself and your affairs.
Jacaranda, Red Lily, Sunshine Wattle, Isopogon.

There are many more facets of numerology which can be found in the many good books on the subject. I particularly recommend *Secrets of the Inner Self* by David A. Phillips, with whom I studied numerology for a number of years.

Kinesiology

Kinesiology puts at our disposal another valuable tool for understanding ourselves better. Stated very simply, kinesiology is a system in which specific muscles are tested in order to provide information about physical and emotional states. With muscle testing we can determine exactly what an organ or an emotional state requires to bring about harmony again. Through the muscles we are able to tap into the nervous system and the brain. Then we can ask the body what it really wants, and virtually anything else we wish to know. We can muscle test to find which bush essence a person needs and how long it should be taken.

Having used muscle testing for many years now, I have come to the conclusion that we can easily rely too much upon this technique instead of following our gut feelings. With muscle testing we are often only confirming our own intuition. I believe that our intuition is one of the most important assets that we have. So, develop your intuition and don't allow its importance to diminish while you acquire other valuable skills.

Nevertheless, kinesiology provides us with another advanced technique that can greatly assist us in our striving to release the limiting factors in our lives and in the lives of our children, born or unborn. It can help individuals lead happier, more fulfilled lives by expressing themselves more powerfully, lovingly and abundantly, thus increasing the chance for greater peace and joy on this planet.

Muscle Testing

A few basic tips:
1. Avoid eye contact with the person you are testing (the subject).
2. Muscle testing is an art; the more often you do it, the more proficient you will become. However, you can achieve clearer responses if you test lightly. Any muscle will give in if tested with enough force. You are not testing muscle strength, just muscle integrity.
3. Don't view the test as a win/lose situation. A successful test is achieved when both people have the desire to locate imbalances so that they can be speedily removed.
4. Keep in mind that the subject is the one in charge. Allow that person to decide when the test is to be done and to provide feedback on whether the testing is too strong or too jerky, etc.
5. Make sure that the subject understands the test and the sequence it will take.
6. Put your expectations aside and avoid trying to influence the results.
7. Do not test if you or the subject have been drinking alcohol or while a television set is switched on near the testing area.

Before testing:
1. Ask for the subject's permission to test the muscle before proceeding, as a muscle may be painful.
2. Switching on stimulates some of the body's major neurological processing centres and allows accurate results to be obtained.

Place one of your hands over the subject's navel, and at the same time rub the points on the subject's lips with your other hand for fifteen seconds. Then repeat the procedure with the opposite hands. The other points to be rubbed are the neck notches. (These are the two bony knobs at the top of the sternal notch.) Again, place one hand on the subject's navel while rubbing the two notches with your other hand for fifteen seconds. Repeat the procedure with the opposite hands.

Alternatively, a single dose of Bush Fuchsia, Crowea or Paw Paw will achieve the same result.

Testing position:
1. The subject stands erect, with feet apart and with the left arm held out at a right angle to the body and parallel to the floor. The subject's limbs should be unclenched and relaxed.
2. The tester stands either in front of or behind the subject, with one hand—the palm flat and facing downwards and the fingers extended—resting on the subject's raised arm, just above the wrist.

    The deltoid muscle is being tested because most people find it easy to reach and it usually locks when tested.
3. The tester demonstrates the range of movement to be performed in the test, so that the subject knows what will happen.

Testing:
1. The subject indicates when he or she is ready to be tested.
2. The tester tells the person to *hold*.
3. With the subject attempting to maintain the pre-test position, the tester applies a constant, gentle but firm downward pressure for two seconds. The aim is to discover whether the muscle "locks", that is, the arm stays up or, if it moves down slightly, whether it springs back up. The pressure should not be so hard that the muscle becomes tired or stressed.
4. If the subject is happy with the test and the muscle locks, a positive response has been achieved. However, if the muscle doesn't lock, that is, the test causes a pain in the arm or the arm shakes, feels spongy or gives way, a negative response is indicated.
5. A quick way to ascertain whether or not the correct response has been given is to ask the subject to state his or her name and then to test the arm. A positive response should result. Then ask the subject to state an incorrect name, and test again. This time you should obtain a negative response. If you do not obtain these results, a dose of Sundew should restore the correct responses.

6. If the subject is unable to give a positive response, the muscle itself may be weak. This can be corrected by rubbing the area between the subject's nipples. Using both your hands with your fingers close together, massage from the breastbone to the nipples.
7. Once a positive response has been established, questions can be asked and the body will give the answers.

The following examples illustrate both the imbalances revealed by muscle testing and the immediately verifiable effects of bush essences in resolving these imbalances.

Stress:

Find a strong indicator muscle on your subject and then ask him or her to think of a situation that is causing a great deal of stress. Retest the muscle while the subject is thinking of the situation. It will usually weaken, indicating that the person is not coping very well with the stress.

Give seven drops of Waratah to the subject and then wait ten seconds or so for the remedy to be absorbed before retesting the muscle. The muscle will usually test strong. Waratah helps individuals cope with major stresses and use their survival skills.

Learning:

Ask the subject to think of a time when he or she felt overwhelmed by a great deal of information, such as at a lecture or seminar. Test the strong indicator muscle while the subject is thinking of that time. It will probably test weak. This time, give seven drops of Paw Paw to the subject, for the assimilation and integration of new information. Wait ten seconds and then retest the muscle. You will usually find that the muscle now locks, showing that the Paw Paw remedy has helped the body utilise the information.

The over-energised person:

You could be excused for thinking that to be over-energised is a positive state, but an over-energised person has great difficulty in focusing and concentrating on any one subject. To determine whether a subject is over-energised, find a strong muscle such as the deltoid. Run your hand lightly up the centre of the person's body from the pubic bone to the chin (your hand doesn't have to touch the subject's body; it can be a few centimetres above it), and then retest. If the muscle no longer locks, the person is over-energised. A dose of Sundew can correct this condition.

One of the benefits of muscle testing is that it brings people in physical contact with one another and encourages healing through caring and touching. If someone feels uncomfortable about being touched, try Flannel Flower, which is a good remedy for assisting people to trust and let go of their inhibitions and to tune into their bodies and enjoy using them.

Touch for Health

This system was devised by an American chiropractor, John Thie, with the goal of making more people familiar with muscle testing, so that with these simple techniques they could improve their own and their families' health.

The basic premise of the Touch for Health system is that certain muscles correspond to vital organs in the body. You can test the impaired functioning of an organ by testing its corresponding muscle, and you can also correct impaired functioning with certain techniques.

There are fourteen main muscles used in Touch for Health, each one corresponding either to a major organ or to one of the two nervous systems. You can tell that an organ, for example, the kidney, is not functioning well if its corresponding indicator muscle does not lock.

A full fourteen-muscle balance, correcting all the major organs and the nervous systems, can take up to half an hour. One dose of Crowea will do exactly the same thing in only a few seconds as it is a very powerful remedy.

With muscle testing you can determine whether a person is suffering from a specific fear and, if so, you can help to clear it. The work being done in this field has been pioneered by Dr Roger Callahan of the United States, the author of the book *Five Minute Phobia Cure*. He discovered that fears are often stored in specific meridians, or energy pathways, of the body, and he devised a technique for their removal.

I have discovered a slight modification that is just as effective but is slightly faster and simpler and uses both flower essences and muscle testing.

Step 1: Find a strong indicator muscle and have the subject state: "I have a fear of . . . " If the muscle doesn't lock, then the body does not have that particular fear. If, however, the muscle does lock, it is affirming that the person does have that fear. In this case, proceed to Steps 2 and 3.

Step 2: Ask the subject to say: "I am now ready to let go of my fear of . . . " If the muscle test is affirmative, that is, the muscle locks, then go on to Step 3. If the muscle is weak, it indicates that the subject is not ready to let go of the fear. To reverse this, the subject pats the outside of the palm, an inch or so below the little finger on one hand at a time, while saying aloud, a dozen or so times: "Even though I am not ready to release this fear, I deeply and profoundly love and accept myself." Then the earlier statement: "I am now ready to let go of my fear of . . . " is retested. If affirmed the person moves to Step 3.

Step 3: Give the subject seven drops of Dog Rose, Emergency Essence or Grey Spider Flower—usually Dog Rose is most appropriate. Then ask the subject to hold an open palm across his or her forehead and think of the situation that would make that fear strongest, until one of two things happens. Either the images will start to fade or the subject will be unable to focus on the fear any longer. One of these outcomes usually occurs within a couple of minutes, rarely more than five. Then ask the subject to restate: "I have a fear of . . . " In most cases, when the subject is retested the fear has been cleared. Note that the same essence used to help clear the fear should be taken for a couple of weeks to reinforce the process.

One of Callahan's primary corrections is for the subject to tap gently under both eyes, approximately thirty-five times, while thinking about the fear. There would certainly be nothing wrong with combining these two techniques, so that if the first wasn't entirely successful, the two together may do the trick.

The way to choose the appropriate remedy—Dog Rose, Emergency Essence or Grey Spider Flower—is to ask the body, using a muscle test. Ask the subject to say: "The most appropriate remedy to clear this fear is [one of the remedies]", and test the muscle. The body will tell you which one it wants.

One final note on Callahan's research: he has discovered that some people consciously want to achieve a particular outcome in their lives, for example, to lose weight, but never do. He claims the reason for this is that they are psychologically reversed, which leads to the subconscious sabotaging of their conscious goals. To determine whether or not this is occurring, find a strong indicator muscle and ask the subject to state: "I want to lose weight." If the person is psychologically reversed, the arm will be weak when tested, yet will be strong after "I don't want to lose weight." Testing "I want to be well/healthy/happy" will also reveal if a person is reversed.

A person who is psychologically reversed usually has a deep dislike of self and will benefit from taking Five Corners. To further reinforce the action of the dose of Five Corners say four or five times aloud, while holding his or her hands over the thymus (breastbone): "I am full of love, trust, faith, gratitude and courage, and I deeply love and accept myself." This affirmation came from John Dramond's work and is very effective. Both the Five Corners and the affirmation can be combined and continued for one to two weeks to reinforce the positive goal.

Age Regression

Of all the techniques I have used in my clinic, age regression stands out for having achieved some of the most amazing results. It is a powerful way in which to clear out deep blocks, traumas and negative beliefs from the subconscious. This technique allows one to go back to the very first time an emotion, for example, guilt, arose and was not dealt with—and then clears it.

To use age regression, the tester places one hand over the subject's forehead and, with the other hand, holds the two prominent lumps, the common integration points, at the base of the skull.

Holding the forehead activates the emotional stress release points, which diffuses stress, bringing calm to a person, and also activates the creative, problem-solving part of the brain—the forebrain. Have you ever noticed that people hold their foreheads when they feel overwhelmed or very upset, and also when they are simply trying to remember something that has slipped their mind or is on the tip of their tongue? We intuitively know what to do even if we can't explain why we do it.

Holding the common integration points helps the mind go back to unresolved emotions in the past. When the mind is given the directive to search for an emotion such as guilt, it will flick through its files and those instances of guilt that have not been dealt with or resolved will spring up in chronological sequence.

Let's take the already-mentioned example of guilt. When the subject is in a comfortable position, either sitting up or lying down, and with the eyes closed, hold the appropriate points on the forehead and at the base of the skull. Then ask the subject to go back to the earliest remembered time when he or she felt guilty and to tell you when an image comes to mind or a feeling arises. Allow the person a little time, thirty seconds or so should be enough, for things to start happening. You can tell when someone is processing those past experiences, searching for the unresolved issue or in fact reviewing that instance because the subject's eyelids will begin to move quite rapidly (REM). If the person reports that an image has come to mind, ask for an estimate of his or her age in the scene. The first image is often a very early memory.

Many people are surprised not only by recalling the scene, but also by the vividness of their memories. They often recall such details as smells from that time or the names of long-forgotten teachers or friends.

As the person relives the memory, be sure to keep your palm over his or her forehead to deactivate any stress. Then, to resolve the incident, you can follow this procedure:

Step 1: Tell the subject to imagine the scene as a piece of celluloid film and himself or herself as a movie director. Then ask the person to chop up the film into many small pieces.

Step 2: Ask the subject, as movie director, to recreate the scene as he or she would like it to be now, so that the scene feels complete.

At this point you may want to suggest a possible scenario. As you touch the person, holding those points, you often become very responsive to and very intuitive about the person's needs. Trust your gut feeling and say what comes into your mind. For example, the person may be processing guilt and may have gone back to the age of three, when he or she was very jealous of a new sibling and treated it very poorly.

After the original memory has been cut up, you could suggest that in the new scene a loving bond exists between the person and the new sibling, a sharing of toys and games and a real enjoyment of playing together. I always add at the end of my suggestion: "But you'll know exactly what you need to do to make it feel complete." And the person does know how to create the scene so that it feels complete!

Step 3: Now ask the person to go back to the next time recalled when he or she experienced guilt. Note that only unresolved experiences of guilt will pop up.

Step 4: Repeat the instructions for chopping up the old memory and recreating the new scene.

Step 5: At the completion of each incident of guilt, ask the person to go back to the next experience until there are no more conscious memories of that emotion. You may be surprised by how frequently scenes arise in which the person is younger than twelve years old.

At the completion of an age regression session, I usually give the subject a dose of Fringed Violet. While reliving those old memories, the person has experienced his or her emotional self very deeply, and the remedy helps to close off the aura and keep it intact. As I've said, it's a very powerful technique and a great deal of emotional garbage can be cleared in a very short period of time.

After a session of age regression, so many people look quite different and much younger. Reinforce its benefits by giving the subject the remedy that deals with the appropriate emotion. In the case of guilt, give the person the Sturt Desert Rose remedy and have him or her take it for one to two weeks afterwards.

A variation on the age regression technique deals not only with emotion. You can also take a person back to a time of great physical trauma which has not yet been resolved in the body. The procedure for this technique is as follows:

Step 1: Using muscle testing, ask the body if unresolved physical trauma is present.

Step 2: If the answer is yes, ask the body to go back to that time and to show when it has regressed to that time with a weak indicator muscle. That is, on muscle testing, you should find that the muscle will not lock, that the arm is weak.

Step 3: At this point ask the person to open his or her legs wide, which simply locks the age when the trauma occurred into circuit. An option is to muscle test to determine the age you are dealing with. This is done by testing whether the age is up to ten; from ten to twenty; from twenty to thirty, etc. The test that is affirmed can then be gone into in detail, and the specific year worked out.

Step 4: Ask if the person remembers the event at that particular time. Most people will.

Step 5: Give the subject a dose of Fringed Violet. Then, with the person's legs still open, test the indicator muscle. It should lock at this point, indicating that the shock and trauma of the event at that age has been cleared by the Fringed Violet essence.

Step 6: Now ask the person to close his or her legs and come back to the present time. Ask the person to state that all stress from that trauma at that age is now totally clear to check whether or not the stress from that earlier time is still effective in the present. Then test the indicator muscle.

Step 7: If the muscle tests weak, give the person a dose of Fringed Violet, and ask him or her to repeat the statement. When you muscle test this time, the stress should be cleared in the present time.

Fringed Violet has the ability to go back to a time of shock and clear the stress from the body. If, for some reason, Fringed Violet doesn't work, repeat the above technique, this time using Emergency Essence—the most powerful remedy for serious physical trauma. In most cases, however, Fringed Violet will be sufficient.

There is much scope for further work to be done with muscle testing, especially when used in conjunction with the Australian Bush Flower Essences. Anyone wishing to find out more about kinesiology should contact the Touch for Health Foundation of Australasia, P.O. Box 164, Buderim, Queensland 4556.

—Affirmations—

I have included this section on affirmations because they can be very useful tools for further empowering the actions of the bush essences. Affirmations can be written, spoken, sung or listened to. They are positive statements which help to program the subconscious mind for particular goals or in particular directions, and they are very easy to use. I have found writing affirmations to be the most effective method, for then you can do two things at once—you can say the affirmations while writing them.

You can use an affirmation anywhere and at any time, though it should be used when you can really focus on it. An affirmation works well if it is used at the same time as taking the corresponding bush essence remedy, that is, either on rising or retiring, or at both times of the day. Writing your affirmation once a day for a week is usually sufficient time for it to work.

If you want to create your own affirmations or to modify the affirmations provided in this book, you can either phrase them so that the statements are in the present ("I am now . . ."), or so that they have a sense of "becoming" ("I am now becoming . . ."). For example, simply writing "I am a loving person" will most likely bring up a lot of resistance in the subconscious mind and will be easily rejected. However, the affirmation "I am now becoming a loving person" will be accepted far more readily by the psyche, which will make the affirmation more powerful.

A highly recommended technique to incorporate into your affirmations is to refer to yourself in the first, second and third person. "I, Ian, am now beginning to love and accept myself" is an example of using the first person in an affirmation. I would write this out ten to twenty times, and after each affirmation I would write my response to it. For example, "If no one else does, why should I?" What the response represents is the negative pattern or belief that is held in the subconscious. As you use affirmations, old garbage from the subconscious is revealed, and by writing out those old messages you are helping to eliminate them.

After writing in the first person, change the affirmation to the second person: "You, Ian, are now beginning to love and accept yourself." Putting the affirmation in the second person will help to clear negative beliefs that have come about from others speaking to you directly. For example, being told, "No one loves you". This leads to the belief, "I am unlovable". After writing these ten or twenty times, go to the third person: "He, Ian, is now beginning . . ." This will cover those beliefs that have arisen from overhearing someone talking about you.

Your response to affirmations can give you valuable insights into some of your unconscious beliefs. You are likely to find that the intensity of their negativity decreases as you write out your affirmations and responses. Sometimes you may get to the point where you basically have no response at all. But, remember, when writing out affirmations do concentrate on them.

At the end of each bush essence entry a couple of affirmations appropriate for that essence have been provided. These have been compiled by Amanda Davey of Melbourne, creator of the Balance for Life program. Russell Sharpe, owner of Just For Love, Sydney's premier florist, and Gina Vanderhage. You can choose an affirmation that you are drawn to, to accompany a specific remedy, but don't feel that you have to use the affirmation in that particular form. If an affirmation doesn't feel totally right, then change its wording. You can tell when you find the form that is right for you.

There are many other ways in which affirmations can be used. They can be listened to on a tape recorder, they can be said while looking in the mirror, or they can be read aloud from the fridge or bathroom door. Take a dose of Turkey Bush, be creative and discover other ways of using them.

The
Bush
—Essences—

SWAMP BANKSIA

(Banksia robur)

*T*he block of granite which was an obstacle in the pathway of the weak, became a stepping-stone in the pathway of the strong.—*Thomas Carlyle.*

The name Banksia honours Sir Joseph Banks (1743–1820), the botanist who came to Australia with Captain Cook on the *Endeavour* in 1770. Banks played a leading role in the colonisation of Australia and was a benefactor of science and President of the Royal Society. The species name *robur* is from the Latin meaning "strong".

Banksias belong to the Proteaceae family, and there are over fifty species in Australia. Although *Banksia robur* grows only up to 3 metres high, this shrub is one of the most striking of all the banksias. Its exceptionally broad, large, leathery leaves are dark green above, with a prominent yellow midrib below, and have irregular and sharply toothed margins. The flower spikes are 8 to 15 centimetres long and 8 to 10 centimetres in diameter. They are a deep bluish-green initially, turning yellow as they open, and are covered in densely packed flowers which run in a spiral pattern up from the base.

With a distribution from the Illawarra region south of Sydney to southern Queensland, this shrub is found on swampy heaths in coastal areas.

This remedy is for situations involving temporary tiredness, frustration or setbacks. It is for those people who are normally very dynamic, with abundant energy and enthusiasm, but who, for some reason or other—illness, disappointment, burnout, etc—are left feeling quite downhearted or flat. A common comment from them is: "I just don't feel as well as I used to." This is an unusual state for such people to be in, one they find not only foreign but extremely exasperating, and they just want to be their old selves again.

Moreover, many people experience the ebb and flow of energy as a natural cycle, and this essence can help them cope with the frustration of the low end of the cycle.

Banksia robur can certainly help these people by lifting their feet out of the bog they're in and getting them back on solid ground. It acts as a wonderful catalyst, and its effects can be greatly enhanced if these people also bathe in fresh water two or three times a day—but not in salt water such as the ocean. This practice, which is peculiar to this essence, helps to wash away negativity. The plant's common name is Swamp Banksia, as it is commonly found growing along creeks, and this seems to explain why washing in fresh water is helpful with this remedy.

I enjoy vitality in all aspects of my life now.
I now feel joy, energy and enthusiasm for life.

Negative condition

- low in energy
- disheartened
- weary
- frustrated

Positive outcome

- enjoyment of life
- energy
- enthusiasm
- interest in life

BAUHINIA

(Lysiphyllum cunninghamii)

The purpose of life after all is to live it, to taste experience to the utmost, to reach out eagerly and without fear for newer and richer experience.—Eleanor Roosevelt

This is a common tree in the tropical woodlands of northern Australia, extending from the plains of the western Kimberleys, south to Port Hedland and east to the Northern Territory. It was previously known as *Bauhinia cunninghamii* and is also called the Bohemia Tree (a corruption of the name Bauhinia).

The species is named after the botanist Allan Cunningham (1791–1839) who was one of the most widely travelled scientific explorers in the history of Australia. He was a protégé of Sir Joseph Banks. The harsh conditions and poor food supplies during his expeditions brought about his death at the age of forty-eight. He left as his legacy a large volume of work on hundreds of Australian plants which he had identified.

Bauhinia can grow up to 10 metres high, though it is usually smaller, and has a short, stout trunk covered by dark grey, fissured bark. The tree has a drooping appearance due to its hanging outer branches. The distinctive large, oval, blue-green leaves consist of a pair of leaflets resembling a butterfly. The tree loses its leaves from June to September when it reveals its small, velvety, orange-red flowers. The flowers produce large quantities of nectar and, along with the fruit pods, are high in protein.

While staying at an Aboriginal cattle station at the top of the Kimberleys I had a dream of an amazing tree. The next day the elder of the tribe was showing me the medicinal plants they used when he led me to the very tree I had dreamed of that night—the Bauhinia tree. I knew instantly that this was to become one of the bush essences.

Geikie Gorge in the Kimberleys was the scene for the making up of this essence. This magnificent gorge cuts through an enormous fossilised coral reef which was formed 350 million years ago when a vast tropical sea covered much of Australia's north-west.

The tree whose flowers I used for the essence was growing on a dry, dusty creek bed. As I tuned into the tree, I received a strong message to hug it. My first reaction was to recoil. Did I really need to hug that rough, gnarled and dirty bark? Yes! came the reply. Oh well, I thought, who am I to argue! As I began to hug the tree I understood Bauhinia's properties in the most wonderful way. The fact that I first had to embrace the tree is so appropriate, for one of the main qualities of Bauhinia is to take in, grasp and accept new ideas and people, even when there is initial reluctance or even dislike.

We live in a technological and information age, in which new ideas are formulated and new information is processed and communicated at a staggering pace. Can you imagine the world at the beginning of this century, when there were no radios, TVs, computers, telephones, or air and space travel. Our world has been transformed by more new technology and information in the last ninety years than in the rest of recorded human history! No longer do we have the luxury of slowly adjusting to and adopting novel concepts—new technology and methods are being used now and are rapidly changing, leaving the antiquated behind! Older managers reluctant to adopt computerisation are now being replaced by younger, computer-literate executives.

New technology and information offer us an exciting period of development unparalleled in our history, provided we are able to cope with the rapid rate of change. Bauhinia can be of great benefit to us in this time of transition.

Due to the mass media, international travel and migration, many people are being exposed to new cultures. As human beings have a tendency to reject or be threatened by anyone or anything different, this is another situation in which Bauhinia proves to be very valuable. It allows us to appreciate people, their customs and circumstances as they are, even if they are new and unfamiliar. In the case of racial hatred or prejudice, the remedy to choose is Slender Rice Flower.

The common name Bohemia, which I discovered after making up the essence, is appropriate for this plant as, in its properties, it is very much like a bohemian, someone who is willing to listen to and accept ideas that are new and quite different from the norm. This remedy provides the opportunity for people to become more flexible in their outlook and more willing to consider new points of view.

Bauhinia can help people come to terms with someone who irritates them. It helps them accept that individual, "warts and all". Of course, a remedy such as Bauhinia can help to

Negative condition

•

resistance to change

•

rigidity

•

reluctance

Positive outcome

•

acceptance

•

open-mindedness

bring greater understanding and less conflict into our rapidly changing world.

One patient mentioned that she found it very difficult to implement the changes to her lifestyle and diet. A change suggested in one consultation would take weeks to bring about. I told her about Bauhinia and she was very excited by its potential.

After a few weeks of taking the essence she reported that it had been instrumental in her decision to learn to use lasers and fax machines in her job as a graphic artist, and that she had made some other major changes at work, too. She found that she was able to handle new situations much more easily and coped much better with the changes in her diet, etc. Even her husband's irregular work schedule didn't upset her as it had previously.

I am now able to embrace new concepts and points of view.
I willingly accept new people and situations now.

BILLY GOAT PLUM

(Planchonia careya)

The body is a temple of God and all parts and aspects of it are perfect.—Kristin White

This is a very widespread plant, found from the northern Kimberleys, through to the Top End of the Northern Territory and across to northern Queensland.

Billy Goat Plum grows as a small, spreading shrub in woodlands, while in moist places, such as monsoonal forests and the edges of flood plains, it occurs as a large tree up to 10 metres high, with thick, rough, grey bark. Its leaves are oval and pale green, and in the late dry season they turn bright red before falling off. The scented flowers are large and fleshy and have long, green calyx lobes and either white or yellow petals. The numerous long stamens arise from a reddish base and are white towards their tips. The flowers appear from July to October.

The Aborigines found many beneficial uses for this plant. The inner bark of the tree, after being pounded and soaked in water until it turned red, was used as a wash for boils, burns and sores. The small, fine roots were soaked in water

and applied to the skin to relieve itching from heat rash, chickenpox or prickly heat.

This bush essence was made up in Kakadu National Park. After making up an essence, I have often discovered during my research that there is a striking and direct connection between its traditional or historical uses and its properties as a flower essence. In the case of Billy Goat Plum, the Aborigines used it a great deal to treat skin problems.

As an essence, Billy Goat Plum is used primarily for people with feelings of self-disgust or self-loathing, especially when this is centred on the sexual organs and/or the sexual act.

A common attitude in our culture is that it doesn't really matter what is happening inside our bodies as long as it doesn't appear on the outside. A rash or elimination through the skin is reacted to with disgust. Most people don't want to be seen with a skin affliction, even though in many cases it is simply the body's attempt to heal itself.

Many people regard herpes and venereal warts with great revulsion, and a discharge caused by thrush or a venereal disease is treated with even greater horror.

As mentioned, Billy Goat Plum can be appropriate if those feelings of self-loathing are a reaction to sex itself. The body secretions, whether perspiration or sexual juices, may be viewed as disgusting, and some people have feelings of such revulsion towards sex that they are unable to enjoy or even participate in the sexual act. Those who have been raped are often left with a sense that their bodies are dirty.

However, the use of Billy Goat Plum is not limited to the sexual sphere but is for any feelings of self-disgust. It can be used topically or taken internally for eczema or psoriasis on any part of the body, provided the disorder elicits a sense of the body being unclean.

Negative condition

•

inability to enjoy sex

•

sexual revulsion

•

physical loathing

Positive outcome

•

sexual pleasure and enjoyment

•

acceptance of the physical body

•

open-mindedness

Of course, the positive side to this remedy is that it helps to bring about a real acceptance of the physical body and the ability to enjoy physical sensations and sexual pleasure. It helps us realise that there is more to people than just their appearance and that if we look deeper we can see true beauty. Of all the Australian flowers, Billy Goat Plum is one of the most spectacularly beautiful.

This remedy was made seemingly by chance, although we know that chance doesn't really exist. Kristin and I had camped for the night in Kakadu, and on awakening in the morning I walked past a flowering Billy Goat Plum 20 metres or so from the tent. As I stood looking at it, the plant conveyed its properties to me and we had a wonderful conversation. To see the beauty of this plant is to understand how it can teach people to see the beauty in themselves and to delight in their physical beings.

I now acknowledge my body with more affection and respect.
I now delight in all aspects of my sexuality.

BLACK-EYED SUSAN

(Tetratheca ericifolia)

*D*on't try to force anything. Let life be a deep let-go. See God opening *millions of flowers every day without forcing the buds.—Bhagwan Shree Rajneesh,* Dying for Enlightenment

Tetratheca has an extensive distribution in the heaths and open forests of the sandstone areas of all States. This genus is one of three in the Tremandraceae family that are native to Australia.

Black-eyed Susan is found on the sandstone plateau that encircles metropolitan Sydney, an area famous for its hardy flora and diversity of species. The area is also known as the Hawkesbury Plateau and is one of three major sandstone areas in southern Australia, the other two being the Grampians in Victoria and the Stirling Range in Western Australia. On the ridges and slopes of these areas and in their valleys some of the most beautiful and rare plants in Australia can be found. The heathlands are the wildflower gardens of Australia.

The soil of heathlands is usually thin, sandy and easily eroded and moisture is scarce, even in high rainfall areas. These adverse conditions are reflected in the stunted growth of the shrubs and the openness of their canopy. Open, dry sclerophyll forests and woodlands can be found on the steep valley sides, with smaller plants growing between the trees. In the Sydney sandstone region boronias, eriostemons, grevilleas, hakeas, epacris, croweas and tetrathecas have evolved.

I grew up in this environment at Terrey Hills. My grandmother would often take me for walks in the area, pointing out the various plants and explaining their properties. Now that I look back over the fifty plants we have used for the bush essences, I can see that, although the flowers were picked from plants all over Australia, many of them can be found in the Sydney sandstone region.

Black-eyed Susan's delicate blossoms grow on small shrubs which are rarely more than 30 centimetres high. The four petals are a rosy mauve colour and the flowers' heads hang down to form a bell-like shape, hence the name *Tetratheca* which means "four-sided box". The plant has whorls of fine, short, woolly leaves. The "black eyes" are the black pollen-bearing tips of the stamens.

As with many of the essences, the properties of Black-eyed Susan are revealed by its structure, which makes it a very good example of the Doctrine of Signatures. I have a very special affinity with this plant, as it is my constitutional remedy.

I would come across this plant regularly in the bush, usually at a time when I was returning to the clinic to see patients. Invariably, I had little time to get back and was running through the bush when I saw the plant. I always felt drawn

Negative condition

•

impatience

•

"on the go"

•

continual expenditure
of energy

•

constant striving

Positive outcome

•

ability to turn inward
and be still

•

slowing down

•

inner peace

to it, torn between wanting to stay with the plant and having to return to the clinic, and this demonstrates the main feature of Black-eyed Susan. It is for people who are trying to do too much, who are always rushing and on the go, with many projects under way at the same time.

The people who benefit from this essence tend to be very fast-moving and quick-thinking. They become very impatient with others if they don't do things as quickly as these people feel they could. A good indication of whether or not this remedy applies to you can be found in the amount of stress in your life. Black-eyed Susan people are constantly trying to cram too many activities into their waking hours. They create stressful situations for themselves and as a result become easily annoyed, impatient and irritable.

Black-eyed Susan is very much an urban remedy for the Speedy Gonzales type. It helps people slow down, turn inwards and find the still centre within themselves, the place where they will find calmness and inner guidance. It helps them come from a centred, balanced place within themselves and cope better in a busy urban environment.

From my own experience, at times when there are great demands and pressures on me, my effectiveness and productivity are greatly enhanced by quiet meditation, which allows creative insights to emerge. Black-eyed Susan has a similar effect.

This essence can also help these people delegate work, for they have great difficulty in allowing others to take on tasks that they feel they can complete so much faster. They like to go at their own speed, without being hindered by others, and they often think more quickly than they can speak. In conversation they will often finish other people's sentences. These people perceive things quickly and get irritated when those around them don't.

Understanding the metaphysical law of "walk, don't run" is of great benefit to the Black-eyed Susan person. With this, you accept that you are in the right place at the right time. Trust this process and give it a try; you may be quite amazed at the results. It is as if the universe gives you a constant flow of green lights. You can prevent a lot of stress by not rushing, by just accepting that where you are and what you are doing are perfect and proceeding at a comfortable pace.

The Black-eyed Susan person hates waiting, as there is never enough time to get everything done. They are usually quite happy people, although when things don't go their way they tend to fly off the handle. They don't like being criticised very much either. They are often travelling here and there, and because of the speed at which they eat, they may suffer from indigestion. I used to pride myself on being able to eat my breakfast of toast and muesli without spilling a drop while driving to my clinic. After studying kinesiology, I realised and could measure the effect that such a lifestyle was having on my digestion.

As well as indigestion, these people may suffer from diarrhoea, as their intestines are in a hurry too. Headaches are also a common pattern, as are muscular tension, back problems and adrenal stress. The degree of frustration experienced by these people can lead to more serious problems such as cancer, strokes and nervous breakdowns unless these stresses are resolved.

For nervous breakdowns, this remedy works in very nicely with macrocarpa (for burnout) and *Banksia robur*, which is for people who get very frustrated when they are sick. In these cases, Black-eyed Susan may need to be used first.

As well as allowing people to slow down and find inner peace, the positive aspects of this essence are a gentleness and sympathy towards others and the patience to listen to others and get to know them. Once these people develop tolerance towards others they can use their quick, intuitive grasp of ideas and situations to help others in practical ways.

When I was checking the validity of my channellings, one of the mediums I worked with burst into fits of laughter as she tuned into Black-eyed Susan. Gradually she regained her composure and told me that her spirit guides had betrayed my secret and were very happy that I had finally found my constitutional remedy.

These days I take Black-eyed Susan about every six months, and each time I need fewer doses to achieve the desired result. But I certainly always feel much better after taking this essence.

Black-eyed Susan is *the* remedy for stress.

I am now able to achieve all that is necessary in a calm and peaceful manner.
I am divinely guided to my Inner Self and I discover my own rhythm.

BLUEBELL

(Wahlenbergia species)

You find true joy and happiness in life when you give and give and go on giving and never count the cost.—Eileen Caddy, The Dawn of Change

The Bluebell belongs to the Campanulaceae family, which in Latin means little bell. It is found mainly in the Northern

Hemisphere and includes the bluebells of Scotland and the English harebells. The Australian representative is from the widespread *Wahlenbergia* genus. There are about eighteen species known in Australia, occurring mostly on the east coast, and they provide endless trouble to botanists trying to establish their identity due to their polymorphic nature.

For our bush essence we used the Native Bluebell which we found growing in the Olgas, or Katajuta as it's known by Aboriginal people, the spiritual and geographical centre of Australia.

The Bluebell is a perennial herb which grows to about half a metre in height and is commonly found on low-lying ground. It has only a few small, soft leaves and its purplish blue flower is supported on a slender stem. Flowering takes place in spring and summer.

This is a remedy to help open the heart. It is for those who feel cut off from their feelings. Emotions are present but they are not expressed, for subconsciously these people are afraid that their feelings of love, joy, etc, are finite or unrenewable.

A common observation in hospital cardiac wards is that after heart surgery, especially triple bypass operations, most people cry a great deal, releasing much emotion that has been blocked for a long time. It has also been noted that the people who recover from heart surgery and maintain good health afterwards are those, especially men, who cry and let out their feelings. Bluebell can be prescribed for people who have blocked off their emotions as it helps to remove the barrier around the heart chakra.

These people may also be afraid to let go of their possessions, as they subconsciously believe that there is just not enough to go around. They are frightened that if they give what they have away, they themselves won't survive. Quite often this belief stems from a past life. The fear in these people often results in a controlled, rigid and forthright manner and appearance, though these characteristics do not have to be present for the essence to be appropriate.

The Bluebell remedy is also good for young children who won't share their toys. It promotes trust in universal abundance, a belief that allows joyful sharing.

The story of the warm fuzzies and cold pricklies illustrates the properties of this essence:

In the days of long ago there was a wonderful land of great joy. The people who lived there loved one another and never got sick or felt unhappy. This greatly upset a wicked witch because no one would buy her spells, so she spread a rumour that if everyone continued to give away their hugs they would use up their supply of love and there would be none left. So people began to hold back their love, only giving their hugs to close family members.

To make up the deficit, the witch started to sell cold pricklies, which could be given instead of warm fuzzies. They didn't feel as good, but there were plenty of them and people wouldn't run out of their own warm fuzzies. Then the people began to get sick and age and there was a lot of unhappiness in the land.

One day a beautiful white witch arrived. The children were drawn to her because they could sense her warmth and love and she freely gave them her hugs. The children responded by giving their hugs to other people, just as freely and without regard for the consequences.

At first people were afraid, wondering what would happen to the children. But soon they realised that the children were glowing with health and happiness, and they, too, started giving their hugs and love. And health and happiness returned to the land.

To quote from Gerald G. Jampolsky's book *Love Is Letting Go of Fear*: "Giving means extending one's love with no conditions, no expectations and no boundaries . . . The giving motivation leads to a sense of inner peace and joy that is unrelated to time."

I am now able to give and receive love in abundance.
I am now transforming the walls of my heart into bridges.
I now open my heart to giving and receiving love.

Negative condition

· emotionally closed

· fear of lack

· greed

· rigidity

Positive outcome

· opens the heart

· belief in abundance

· universal trust

· joyful sharing

BORONIA

(Boronia ledifolia)

*L*ife *is full and overflowing with the new. But it is necessary to empty out the old to make room for the new to enter.—Eileen Caddy,*
Footprints on the Path

There are over ninety different species in the *Boronia* genus, all of which are found in Australia, with the majority being native to Queensland and New South Wales. The very large plant family to which they belong, Rutaceae, is widespread around the world and includes oranges, lemons, limes, cumquats and many other plants with aromatic properties.

Boronia ledifolia grows as far south as the southern tip of Victoria, and as far north as southern Queensland, being prevalent on the coastal plains and tablelands. It emanates, depending on your taste, either a sweet or pungent smell when its highly aromatic, dark green leaves are crushed.

Boronia was named after Francis Borone, an Italian botanist who died in Athens at the age of only twenty-six.

Boronia ledifolia, like all boronias, has starry flowers with four petals and eight stamens. The pink flowers are 5 to 10 millimetres across and bloom from July onwards, bringing the dry, rocky gullies and hillsides alive with colour. The bushy shrub, which is up to a metre tall, grows in semi-shaded areas of dry eucalypt forests in sandstone regions.

After the flowers have bloomed they close like buds over the ripening fruit. The seed, which is within a capsule, is explosively scattered when ripe.

The Boronia essence has two main properties. The first leads to clarity of thought and serenity of mind. It is a wonderful remedy for those who have obsessive thoughts going round and round in their minds, which they are unable to release. They may keep thinking about and reviewing a situation or conversation. This essence resolves thoughts that are stuck. Like the seeds, these unwanted thoughts are explosively flung from the mind.

This remedy helps to quieten the mind, which in turn allows the intuitive faculties to function more completely. This dynamic stillness of mind is also experienced in meditation. The repetition of a mantra is often used to give the mind something to focus on in order to release its chatter. This is also the effect of taking the Boronia essence.

Some people experience insomnia because of obsessive thoughts. Boronia can help to turn off the dialogue and clear the mind so that they can sleep. It also helps people discard unpleasant thoughts. This remedy was prescribed for an eighty-three-year-old woman who had been hearing singing voices at night, which disrupted her sleep. After taking the essence she was sleeping better and her mind was more balanced.

Boronia people often have a feeling of fullness or pressure in their heads and find it difficult to concentrate. Because they are not focused purely in the present, they may be accident-prone. These people take their work home with them at the end of the day.

The positive aspects of the Boronia remedy allow creative visualisation to be used and focused in a very powerful way. This technique helps people manifest desired changes in their lives and heal their own bodies.

The second area over which the Boronia essence exerts a major influence concerns pining for another person. It can be taken immediately after a relationship has broken up and all one's thoughts are focused on the lost partner, particularly when one feels a lot of hurt and sadness. In this situation it combines very well with Bottlebrush, which allows one to release the past and move forward. For grief or sadness of long duration, use Sturt Desert Pea.

I now release others to be who they are.
I now release persistent, unwanted thoughts and replace them with inner peace and serenity.

Negative condition

•

obsessive thoughts

•

pining

•

broken-hearted

Positive outcome

•

clarity of mind and thought

•

serenity

•

mental calmness

BOTTLEBRUSH

(Callistemon linearis)

Every end is a new beginning.—Begin It Now *(ed. Susan Hayward)*

Callistemon is a genus in the Myrtaceae family consisting of small trees and shrubs. Of the twenty-odd species, sixteen, including *Callistemon linearis*, are found in New South Wales.

The generic name *Callistemon* means beautiful stamen, and the showy bottlebrushes are made up of many individual flowers packed in dense cylindrical spikes on the tips of the branches. The colour lies in the numerous stamens which are much longer than the five inconspicuous petals. The bottlebrushes, abundant in late spring, are about 10 centimetres long and a brilliant red. A new growth of silky shoots begins in the apex of the flower spike, and in the following year a new spike will form at the growing tip.

The fruit capsules from previous years remain clustered in solid masses of successive groupings lower down on the branches, indicating the age of the plant.

Preferring damp ground, the Bottlebrush can grow up to 3 metres tall and has long, stiff, narrow, dark green leaves.

The bottlebrush essence is for major transitions in life. It helps to give people a belief in their own ability to handle new situations.

Before making up this essence I was working with a number of pregnant women. I became aware of the need for a remedy that would help them deal with feeling overwhelmed by their physical condition and by the responsibilities to come. Many of them felt inadequate at times.

The information on Bottlebrush was channelled to me at this time and, appropriately enough, its essence is made not only with the flowers but also with the fluffy new leaves which are extremely soft like the skin of a newborn baby.

This essence aids the bonding between a mother and her child, which may be hindered, among other things, by a mother's negative feelings, especially during her first pregnancy. Pregnancy is a good time for clearing out a lot of emotional garbage which could adversely affect the child.

A number of other physical changes occur during a person's life, including growth, lactation, menopause and even death. There are also natural cycles of change in life. Every seven years all our body cells are replaced and, metaphysically, every

seven years up to the age of twenty-one another outer body is formed. At twenty-one, when the astral and etheric bodies have been formed, a person has a good chance of living to old age, which is why we traditionally celebrate that birthday.

This remedy can be used by dying people who have a spiritual awareness and understanding of death and the afterlife. It will help them deal with other people's expectations, as well as allowing them to come to terms with dying.

Of course, if we look past these physical changes there are many other major turning points and transitions in life. Starting school, getting a job, marriage, divorce, moving house and retirement all may involve feelings of uncertainty and apprehension. These are times when Bottlebrush can be very helpful.

Another aspect of this essence is that it helps to brush away the past and allows a person to move on to new situations and experiences. Life is full of ends and beginnings, bringing constant change. To resist change is to block the flow of life.

If you look at your life and realise that you are not making any new friends or meeting new people, that you always drive the same way to work and rarely change your hairstyle or the type of clothes you wear, then Bottlebrush may be the catalyst you need to get the life force flowing again.

People close to death can take it twice a day or as often as remembered. The remedy is usually quick-acting.

Negative condition

•

overwhelmed by
major life changes

•

adolescence

•

parenthood

•

pregnancy

•

old age

•

approaching death

Positive outcome

•

serenity and calm

•

ability to cope

•

ability to move on

It is not advisable to give this remedy to children under twelve years of age as, metaphysically speaking, they chose the events in their lives up to that age before they were born. Although Bottlebrush wouldn't be harmful, it would be better if they were allowed to experience those events without outside interference.

The positive nature of this essence allows a person to flow through life and its changes, being able to let go of the past and move ahead into new experiences.

I now break all links that have been hindering my growth.
I now move easily with life's changes.
I now welcome the new responsibilities my pregnancy will bring.

BUSH FUCHSIA

(Epacris longiflora)

I learned that nothing is impossible when we follow our inner guidance, even when its direction may threaten us by reversing our usual logic.— Gerald G. Jampolsky, Love is Letting Go of Fear

With the exception of Western Australia and the Northern Territory, the *Epacris* genus is found in all States of Australia. In all, there are approximately forty species in Australia, and one of those, common heath, is the floral emblem of Victoria.

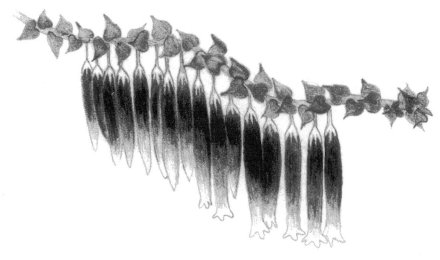

Epacris longiflora, a low, straggling shrub with slender, wiry stems and small, heart-shaped leaves, is the one used for the bush essence. It is one of many plants classified as sclerophylls, or hard-leafed plants, which all exhibit methods evolved to compensate for the usually infertile soil and hot, arid environment in which they grow.

Bush Fuchsia's long, red, tubelike flowers, tipped with white, create a brilliant show of colour all through the year but particularly in spring, lighting up Sydney's sandstone ledges. Not being as light-demanding as other species, it can be found among the undergrowth of heathlands and dry sclerophyll forests along the New South Wales coast and Great Dividing Range.

The channelling of this energy was a collaborative process between me and Kristin. I channelled down the negative aspects of the essence, while Kristin got those for the positive outcome.

Bush Fuchsia has a number of applications that are especially important for learning problems, as it is able to integrate the left and right hemispheres of the brain and most learning difficulties stem from imbalances between these hemispheres. It can be taken for dyslexia and has been tremendously beneficial to many people affected by this condition.

I recently received a phone call from some parents whose eight-year-old son had just taken his first dose of the Bush Fuchsia essence. Within minutes of taking it, he had been able to put together a complete sentence for the first time in his life.

The beauty of the Bush Fuchsia essence, like many other bush essences, is its speed in clearing the blocks. An immediate change is apparent after just a single dose of this remedy. At the workshops we give, volunteers read out aloud, and then again after taking a dose of Bush Fuchsia, and the difference is startling. A flat monotone voice suddenly comes to life with inflections and a confident reading style. A number of institutions and teachers working with people with learning difficulties are now incorporating this essence into their programs.

After taking a single dose of this remedy, a seven-year-old girl who rarely spoke didn't stop talking for three days! By the fourth day her conversation was normal.

Other indications for this remedy can be an inability to read for very long because one falls asleep or loses concentration. What is happening is that the mind isn't integrating the information that the eyes are perceiving, and Bush Fuchsia essence will correct this. Children who never read will begin to ask their parents and teachers for books after taking the Bush Fuchsia essence.

For students who battle through lessons, the Bush Fuchsia

Negative condition

•

dyslexia

•

poor learning ability

•

stuttering

•

nervousness in public

•

ignoring "gut feelings"

Positive outcome

•

courage to speak out

•

clarity in speaking

•

in touch with intuition

•

balancing and integration of left and right hemispheres

remedy makes school an exciting rather than a depressing place. Most children who have problems with maths, for example, switch off in primary school and never catch up. As well as making learning fun, this remedy increases students' confidence among their peers in the classroom.

This essence balances the left side of the brain (logical, rational functions) with the right side of the brain (intuitive, creative functions). It is excellent for helping people get in touch with their intuitions and, even more important, trust and listen to them. Ninety-nine per cent of the time our gut feelings are right, but in our left-hemisphere-dominated society our reaction is often to consider these intuitions in a logical fashion and then dismiss them. The great scientist Albert Einstein stated: "I never came upon any of my discoveries through the process of rational thinking."

I had a very interesting experience one day while driving across town to a workshop. As I was going over the Harbour Bridge my gut feeling was to count how much money I had in my wallet. I dismissed this as merely a sign of avarice. At lunchtime I went to my bag to get my money and found my wallet was not there. For the rest of that afternoon I had to contend with gnawing, worrying feelings about whether it had been stolen or I had left it at home. It contained the rent money, too, that day. On going home, I found that I had left it by my bed, but if I had trusted my gut feeling that morning I would have been saved those hours of anxiety.

There are many times when, if we listened to our gut feelings, our lives would be much simpler. I know a number of women who, when pregnant with their first child, were being influenced by other people and didn't have the confidence to follow their own gut feelings because they were concerned that these might not be right for their babies. Their whole experience of motherhood changed once they took this remedy and trusted and followed their gut feelings.

For people who spend a lot of time in front of videos or electrical equipment and feel a bit dull or "switched off" at the end of the day, Bush Fuchsia will switch them back on again. You can often tell when you are in this state because you start to make silly mistakes.

Bush Fuchsia increases the clarity of speech and is helpful for stutterers. It will give people the confidence to speak in public as well as the ability to speak out about their own convictions and to express what they want to say. If a person is very nervous about public speaking, Bush Fuchsia should be taken a few days before the occasion and just before speaking as well. Several actors are now using Bush Fuchsia whenever they go to auditions as they find that they are able to project much more successfully. They report that they are getting roles which they would never have been given in the past.

I have taken this remedy before being interviewed on radio and television and have found it much easier to communicate my thoughts.

For dyslexia, Bush Fuchsia can be taken for a fortnight and then, after a break of a couple of weeks, this can be repeated. It may have to be used for a number of weeks or even months, but once a marked improvement has been brought about, it can be gradually phased out.

I now trust and follow my intuition.
My true voice flows freely, clearly and effortlessly now.

BUSH GARDENIA

(Gardenia megasperma)

The fruit of all good marriages is lasting love. —Kama Sutra

Bush Gardenia is commonly found throughout tropical woodlands and open forests in the Northern Territory. A tree up to 9 metres high, it has a rounded crown and crooked branches, with smooth, mottled, yellowish, powdery bark. The leaves are rounded and leathery, with a prominent herringbone vein, but they are velvety when young. The fragrant white flowers are 4 to 5 centimetres in diameter, with nine petals, and bloom from July to November.

The gardenia takes its name from the American obstetrician and naturalist Dr Alexander Gardener (1730-1791). The species name is derived from the Greek words *mega*, meaning big, and *sperma*, seed. The hard fruit is about 6 centimetres long and contains many edible seeds which are filled with a sweet, thick juice and are embedded in a pulpy material.

The story of the making of the Bush Gardenia essence has been told earlier in this book (see "A Bush Essence Journey"). This remedy is for renewing passion and interest within relationships. It helps to draw together couples who are moving away from one another because they are too preoccupied by their own lives. It is as if the essence helps to turn the individuals' heads to face one another and see what the other partner is doing and feeling and what is needed to bring them back together.

A couple in their eighties who consulted me were growing apart. The husband was offhand towards his wife, and sometimes even rude to her. After he had taken the Bush Gardenia essence, his wife reported that he was bringing her breakfast in bed and telling her that he loved her.

Bush Gardenia is not only for male-female relationships, but also for family bonds. It can be used when a family member is going off the rails because of drugs or another problem, while the rest of the family is unaware of this as they are caught up in their own lives.

A health practitioner has this to say about the Bush Gardenia essence: "I have found that this remedy is particularly appropriate for people who need to have a more loving relationship with themselves, as well as for relationships between young brothers and sisters."

There may be some significance in the nine petals of the flower as in numerology the number 9 represents humanitarian concerns.

The book entitled *How to Make Love to the Same Person for the Rest of Your Life* by Dagmar O'Connor has a similar theme to that of the Bush Gardenia essence. But Bush Gardenia also involves improving communication between couples, as well as between parents and their children.

Of all the flower scents, I can think of none more sensuous than a gardenia's, so how appropriate that this flower should be related to passion and sexuality.

I now create time to communicate with my loved ones with sensitivity and understanding.
I now discover new ways to strengthen my relationship with my partner/child/parent.

Negative condition

•

stale relationships

•

self-interest

•

unaware

Positive outcome

•

passion

•

renews interest in partner

•

improves communication

Bush Iris

(Patersonia longifolia)

Be still, and know that I am God.—Psalms 46:9.

The three purple flags, or sepals, of this attractive flower herald the profusion of spring flowers in the sandstone country of eastern Australia. The sepals provide the flower's colour as the petals are inconspicuous. Though they appear in great numbers, the flowers of Bush Iris are very delicate and bloom for only a few hours. Affected by the heat of the midday sun, they usually fade by late afternoon and then disappear. Yet they can seemingly reappear within a short time. The reason for this lies in the fact that each flower is only one of several which are tightly packed within the same thick, dense bract. So as one dies it is quickly replaced by another opening bud. To the casual observer, it appears that the flower of this small herb has been resurrected from its withered form.

Belonging to the Iris family, the *Patersonia* genus consists of approximately twenty species. All are exclusive to Australia except three that are found as far north as the Philippines. The generic name honours Captain William Paterson, an ardent botanic collector who was Governor of New South Wales from 1794 to 1795.

This is the first remedy that Kristin and I made up. One day we went into the bush to make up the Waratah essence but it wasn't in perfect bloom. Then we found the Bush Iris. There are no accidents! That night in meditation both Kristin and I learned that it was very important for this to be our first remedy, as it would open up the door to a person's higher perception and spirituality, allowing the Trinity to flow into the person.

For Kristin, the wonderful array of mauve and purple blooms provided a psychic doorway to an understanding of the properties of the other plants. This remedy aids spiritual development and allows entry to another level of awareness. It is excellent when taken before meditating or before embarking upon a course of meditation or any other spiritual or religious practice. It enhances self-awareness and makes creative visualisation during meditation much more effective.

A number of metaphysical teachers in Australia give Bush Iris to their students to enhance their spiritual awareness. It helps to clear blocks in the base chakra and in the trust centre, or throat chakra, as well as the third eye and crown chakras.

On the other hand, the negative aspect of Bush Iris is a materialistic, atheistic approach to life—the sex, drugs and rock 'n' roll syndrome—where there is a denial of the spiritual. If a person is stuck in a materialistic mode, Bush Iris can bring balance to their lives. It is also for those who want to become aware of the spiritual realm.

It has been said that for those who believe in God, no proof is necessary, while for those who don't believe, no proof is possible. Yet Bush Iris fosters faith and helps people step forward fearlessly in life, knowing that God is with them.

Bush Iris is also very helpful for the dying. The fear of death leads to a grim holding on to life, which in turn leads to much more pain and discomfort. And for many people, dying is a time of great fear and uncertainty about what will happen. Bush Iris certainly won't kill anyone before their time or before they are ready to go. However, it will make their passing over smoother and more peaceful and will remove much of the anguish and fear often associated with this transition.

In many cases, giving the Bush Iris essence to a dying person can decrease the need for pain-relieving drugs. Those who work with the dying have noticed that, a day or so before they die, people often become very calm. Many, in fact, see

Negative condition

·

fear of death

·

materialism

·

atheism

·

sexual excess

·

avarice

Positive outcome

·

awakening of spirituality

·

assists dying in their transition

·

clearing of blocks in the base chakra and trust centre

relatives or loved ones who have already passed over, as if their spirits have come to guide the dying person through the Light. However, the drugs given to the dying, such as pain-killers to cancer patients, can block their psychic perception and thus deprive them of these important experiences.

I have prescribed this remedy a number of times for patients who are close to death. One was a Vietnam veteran who was so afraid of dying that he wouldn't let anyone, not even his children, talk about his illness or mention death. A great amount of tension built up in the household because everyone knew he was dying but were afraid to talk about it. When he took Bush Iris the family was finally able to discuss their feelings openly before he died. This was a great relief to them all, and after he died his wife thanked me for recommending this essence, which had allowed her husband to experience such a peaceful death.

I am now attuned to my awakening spirituality.
I now accept death as only another change.
The Universe always protects me.

CROWEA

(Crowea saligna)

I'm an old man and have known a great many troubles, but most of them have never happened.—Mark Twain

Crowea is a small shrub, up to a metre high, which is particularly common along the sandstone ridges of the coastal regions of New South Wales and southern Queensland.

The genus derives its name from the English surgeon J. Crowe (1750–1807) who did a considerable amount of research into mosses and fungi. Flowering from autumn to early spring, Crowea's five magenta-pink petals form an open flower which is 3 to 4 centimetres wide. The long appendages on the stamen tips give the flower a very prominent centre. A striking contrast to the flowers is provided by the dark green, lance-shaped leaves. It's lovely to come across a flowering Crowea plant in the bush for it has a very regal presence.

Crowea is a frequently prescribed remedy which has immense value. It has a strengthening, calming and centring

effect on the body and mind, invariably bringing about a tremendous sense of well-being and aliveness. For those people who don't feel quite right or simply feel out of sorts, Crowea is the perfect choice as it helps them find their centre again. Often they are not able to put their finger on exactly what is wrong.

Recently I travelled to Brisbane to run a workshop on the bush essences, and as I arrived on the first morning I found the venue locked and the caretaker away for the weekend. After trying every window, balcony and door for a way in, I came across a skylight on the roof which I was able to prise open. I then had to jump down onto a washbasin in one of the bathrooms. After finally getting in without breaking any bones, I opened the front door to bring in the equipment, which still had to be set up, and the participants began to arrive. Hardly the best way to start a day of teaching!

A single dose of Crowea helped me to relax, eased my sense of panic and allowed me to enjoy the workshop. It is a great remedy for worry and stress.

Crowea is an excellent remedy for stomach ulcers and other stomach ailments. The five petals of the Crowea flower are numerologically linked with emotional balance and the stomach.

I have made a very interesting discovery about this plant's relationship to kinesiology (for further details on this subject, see "Kinesiology"), especially the form called Touch for Health, where specific muscles indicate how a corresponding organ, for example, the stomach, is functioning. To do a full Touch for Health "balance" on the fourteen major organs and nervous systems takes an average of 20 to 30 minutes of physical

Negative condition

•

continual worrying

•

feeling "not quite right"

Positive outcome

•

peace and calm

•

vitality

•

balances and centres the individual

work. Taking just seven drops of Crowea—one dose—will give exactly the same result, and the effect is just as long-lasting.

This remedy will also help to align the etheric and astral bodies with the physical body when these are out of balance, as can occur when a person is physically disturbed whilst astral travelling, such as being rudely awakened from deep sleep.

Hurry, worry and money have been described as the three evils of humanity. If this is so, then the Crowea remedy addresses a third of the evils that beset us, for it is a great remedy for worry. Ninety per cent or more of what we worry about either never occurs or else is something that we have no control over, so we may as well not waste time dwelling on those possibilities. My own case histories as well as feedback from others show that Crowea is excellent for incessant worriers.

I am now releasing worries and opening myself to more vitality and inner peace.
Peace of mind is mine now.

DAGGER HAKEA

(Hakea teretifolia)

To forgive is the highest, most beautiful form of love. In return you will receive untold peace and happiness.—Robert Muller in A Bag of Jewels *(eds Susan Hayward & Malcolm Cohan)*

This genus was named after an eighteenth-century patron of botany, Baron Von Hake. There are over 100 species of *Hakea* in Australia, which belong to the same family as the Banksia and the Waratah, namely, Proteaceae. Although most are shrubs, some Hakeas do grow to the size of a small tree. All of the species have thick, often large, woody fruit enclosing two black winged seeds which are released on the death of either the branch or the whole plant.

Dagger Hakea is a straggly shrub which can be found in a variety of shapes but is rarely more than 3 metres high. The white to cream flowers, which bloom in late spring, have one of the most beautiful perfumes of any Australian wildflower, the scent being akin to a sweet honey-cinnamon blend. The fruit, which remains on the tree for a long time, is narrow and pointed like a dagger. It is easy to understand how this shrub got its common name when you see—and feel—both its fruit and its needlelike leaves, which are 2 to 5 centimetres long.

Dagger Hakea is widespread in damp heathland areas and is often found growing along bush tracks, requiring a precarious and often painful passage. The Doctrine of Signatures is very evident with this bush!

Not surprisingly, this remedy is used for those who are a bit "prickly" and whose words can be sharp barbs at times. These people may hold old grudges against those to whom they have been very close, such as family members and past lovers.

This remedy primarily brings about the open expression of feelings and forgiveness. It helps people work through and resolve intense feelings of resentment and bitterness, which are usually directed towards those who have been very close to them. On the other hand, Mountain Devil is for people who experience those emotions towards other people generally.

With Dagger Hakea, resentment is more covert than overt. At times these people's feelings are so intense that they are overwhelming. So they keep their feelings of resentment locked inside. Sometimes these people seem very sweet-natured, but there is a lot of anger behind the facade.

This is a wonderful remedy when a relationship has ended or a family feud has developed. In the former case, before people can release a former lover or friend, they may have to work through their feelings, especially if the other person ended the relationship. A lot of resentment and bitterness can build up if people feel that they have been unfairly treated.

The following process is very beneficial for people wanting to work through and resolve their resentment and bitterness. It is called the Forgiveness Process, and even greater results can be achieved if the person takes Dagger Hakea for up to a week before doing the process, and for a few days afterwards.

The Forgiveness Process

To do this process, it is advisable to find a quiet, private place where you can talk aloud without the fear of being overheard or disturbed. The process can be done over a period of time or it can be done in one sitting, but the latter may take quite a number of hours—so be warned.

After sitting or lying down comfortably, take a few deep breaths and close your eyes. Allow yourself to fall into a relaxed and reflective state. Then ask to be presented, in a visual form, with the people you hold resentment towards. Only one person at a time will appear, and the ones with whom the greatest number of issues need to be resolved will turn up first—almost invariably, first one parent and then the other.

When you see a particular person or just have a sense of who that person is, visualise a cord coming from that person's navel and another cord from your navel. Tie the cords together. Then, speaking aloud, say the person's name followed by: "The resentment I hold against you for . . . " and state everything for which you feel resentful or bitter towards that person.

When you have stated everything, say the person's name and: "The resentment I hold against you I now release. I love you and I forgive you." As you say "I forgive you", make a pair of scissors with your hand and cut through that imaginary cord connecting you with the person.

Repeat this whole process twice, and try to state resentments that you may have forgotten the first time as well as ones mentioned previously.

After completing the two additional processes, you then change the process. This is done by tying the cords between you and the person and this time saying the person's name and: "The resentment *you* hold against *me* for . . . " Then state the things for which you feel that person feels resentful towards you. On completion, say the person's name again and: "All these things for which *you* hold resentment against

Negative condition

•

resentment

•

bitterness towards
close family, friends
and lovers

Positive outcome

•

forgiveness

•

open expression of
feelings

me I now forgive you for. I love you and I release you," and cut the cord. As before, repeat this twice.

You are then ready to embark upon resolving issues with the next person, so at this point ask to be presented with another person you hold resentment towards.

The first time I did this process I came up with quite a few old incidents which had been locked away in my subconscious. In fact, the school bully from over twenty years ago popped up. He was someone I hadn't thought about for years and years, yet the resentment was still there, in my subconscious mind, draining my energy and vitality.

When the whole Forgiveness Process has been completed, you will feel a sense of great relief and lightness.

Betty, aged forty-two, presented at my clinic with gallstones and gallstone colic. She was in her second marriage and admitted that she felt resentment towards her very demanding eight-year-old stepdaughter who lived with them. She was also angry at the amount of time her husband gave to his daughter, yet he spent little time with Betty's seventeen-year-old daughter who lived in the same house.

The situation had come to a head when Betty's husband had made out his will, leaving many household items to his daughter. Betty had actually paid for those articles, but when she had brought this up with her husband, he had still refused to change his will. Betty had kept in a lot of the resentment and bitterness she felt about the situation, which had produced the gallstones.

During the consultation she agreed that Dagger Hakea was most appropriate, and when she returned a fortnight later there had been no recurrence of the pain. In Chinese medicine, the gall bladder and liver are related to anger. She also mentioned that she felt less angry and resentful and was starting to talk about her feelings with her husband. And, as a result, he had changed his will.

She was aware, too, that she was far more tolerant towards her stepdaughter, claiming that the girl had also made some major shifts and was becoming less selfish. It was interesting to speculate about the possibility that the stepdaughter had stayed the same, but Betty was seeing her from a different perspective.

There is a wonderful naturopathic treatment for gallstones which I use quite regularly in my practice. I also prescribe Dagger Hakea to unblock the emotion that is causing the formation of the stones, and the two treatments together are very effective.

I am now able to express my feelings honestly and openly.
I now release all buried resentment and bitterness and am learning to forgive.

DOG ROSE

(Bauera rubioides)

Man cannot discover new oceans until he has the courage to lose sight of the shore.—Anon

The name of the *Bauera* genus honours the memory of Ferdinand Bauer (1760–1826), the botanical artist who, under Joseph Banks' directive, sailed with Matthew Flinders during his amazing circumnavigation of Australia in 1803. As far as botany was concerned, this voyage was the most important ever undertaken up until that time. Bauer was hailed as one of the finest artists in the history of botanic art. His legacy was a magnificent collection of drawings of Australian plants—over 1500 of them—and animals.

Dog Rose is a wiry shrub which can grow up to 2 metres high and is found in all States except Western Australia and the Northern Territory. This genus is native to Australia.

Bauera rubioides has small leaves, and its flowers, with six petals, range in colour from deep, rich pink to white. It flowers from late spring to summer.

Dog Rose grows abundantly in moist coastal forests, along creeks and the edges of wet heathlands and even against wet rock faces, but always near water, and it sometimes forms dense thickets.

Interestingly enough, in Chinese medicine the element of water is represented by the kidney meridian. The emotion associated with the kidneys is that of fear, which is the main emotional state resolved by Dog Rose. The rich pink flowers have a hanging, limp appearance, like the drooping, rounded shoulders of a defeated person or someone low in energy. Dog Rose treats common fears, not terror, but niggling minor fears. Some examples are the fear of being robbed, the fear of swimming and the fear of heights.

Yet fear is a sign of too much concern with self-interest. Energy is directed inwards in a rather morbid way. Being fearful has the effect of blocking and suppressing the life force, the vital energy. Fearful people do not utilise the energy available to them and lack vitality.

Fear also stops love coming in, and it is love that will help to dissipate fear. The pink of the Dog Rose flower is important, as this colour helps to release the love inherent in a person and to open the person to the love around them. Giving and accepting love will help the person let go of fear. Those interested in following up on this subject may like to read Gerald G. Jampolsky's book *Love Is Letting Go of Fear*.

Of course, the overcoming of fear leads to an increase in energy as the life force is allowed to flow. It also increases the quality of life by giving people a sense of their own self-worth. Thus Dog Rose can also be used by shy people if their shyness stems from a lack of confidence with others, rather than from inherent low self-esteem for which Five Corners is more appropriate. Dog Rose gives these people more self-confidence and allows them to feel comfortable and express themselves freely with others. It helps them enjoy other people, rather than being fearful of them.

The consequences of fear on a physical level are, for many people, stomach problems. They may get ulcers due to the high acidity of the gastric juices. Constant fear may also weaken the adrenal glands and lower the amount of oxygen in the body. In iridology, the whitening of the iris and enlargement of the pupil are signs of fear.

As already mentioned, in the Chinese system fear affects the kidneys, which are connected to the bladder. Many children who wet the bed can be helped if the fears that they invariably have are alleviated. They may be afraid of a parent, commonly the father. A child being afraid of the dark indicates that the kidneys are not functioning well. Again, if a child's feet are turned out, splayed like a ballerina's, the kidneys are not functioning properly and this child is likely to be fearful.

The more we focus on a fear, whether it is a vague, general fear or something specific, the more likely it is that our minds will create the very events that we fear. There is an old saying: "Man with forty locks on door attracts forty thieves." It is

Negative condition

•

fearful

•

shy

•

insecure

•

apprehensive with other people

•

niggling fears

Positive outcome

•

confidence

•

belief in self

•

courage

•

love of life

one thing to take care of our possessions, and another to be paranoid about losing them and constantly focusing on the fear that they will be stolen. People who have been robbed can focus so much on the event that they attract other robberies to themselves. Nothing happens to us without a reason, and if we can learn from one event then we don't have to have the same lesson repeated over and over again by the Universe, until we finally understand it. The penalty for being slow to learn is that each repetition of a lesson is presented in a slightly stronger way, the intention being to help us learn it this time.

If you find that you are dwelling on a fear, a helpful technique is to visualise a large black cross going through your negative thoughts and then to imagine a positive version of the scene. You are showing your mind what you don't want and replacing it with what you do want.

The Dog Rose remedy is also very good for people who have bad nightmares or for those who become very nervous after watching horror films or reading thrillers. Fear can be picked up from others or from the general environment. Or an event can trigger off a fear hidden in the subconscious.

This essence can help us learn a valuable lesson. We are able to move ahead in our lives even when we are afraid. We can face our fears and use the extra adrenalin they produce in order to go through fearful experiences. In personal growth courses the expression "to ride the wild tiger" means to confront our fears. Fear is only in our minds, and in most cases the things we fear turn out to be paper tigers.

The section of this book entitled "Kinesiology" discusses muscle testing for phobias. If a person does have a phobia, they can get very good results by using the technique described in that section as well as taking the Dog Rose essence.

Another property of fear is that it attracts negativity and dark forces. These dark forces are very real and, if present, may greatly affect people who feel fearful. The Dog Rose and Grey Spider Flower remedies help them overcome fear in these circumstances and have faith in themselves, in God and the protection of the Light.

Some people are very reluctant to admit that they have minor, niggling fears because they feel that those fears are foolish. These people may be shy and insecure about themselves, and Dog Rose will help them resolve those problems.

Dog Rose can also be appropriate for those who have experienced grief, as unresolved grief often makes people fearful. Boronia and Sturt Desert Pea are the remedies that help people work through their grief, and Dog Rose can then be used if fear is a problem.

Insomnia is quite common in fearful people because they often dwell on their fears. They may also be compulsive eaters or may eat to block out uncomfortable feelings. If they have difficulty in mixing with others, they may be found at home watching television and overeating.

Dog Rose is very useful for wild animals that have been hurt and are being nursed, because of their natural fear of human beings.

It is also a good remedy for people who have a fear of getting a specific illness, such as AIDS or cancer. If you keep thinking and worrying about a disease, you often end up getting it. These fears are often created by the media or by medical authorities. They quote statistics on the incidence of various diseases in the population and advise the public to be screened for a multitude of ailments. This has led to people losing confidence in their ability to deal with their own lives and health. Children can sense a feeling of inadequacy or fear in their parents when they are unable to deal confidently with their families' health problems.

A few generations ago women looked after the health of their own families, to a great extent. In most Western countries today, the knowledge and understanding of healing has been lost by women, as medical authorities have convinced them, often through fear, that only competent doctors with years of medical training can be trusted with their families' health.

The positive aspect of this bush flower essence is an adventurous attitude, being prepared to take on life and being unafraid of new challenges. Life becomes joyful when you are able to meet and enjoy other people and to do what you have always wanted to do, without fear stopping you.

I acknowledge my own magnificence and feel confident in all situations now.
I now enjoy the company of others.

FIVE CORNERS

(Styphelia triflora)

Life is getting to know you are worth it.—*Gita Bellin,* Amazing Grace

The fifteen *Styphelia* species are found in all States of Australia. Five Corners is an erect shrub growing up to 2 metres high. It has flat, very spiky leaves along its stems. The tubular

flowers are a gorgeous pink, and as they develop the five yellow lobes roll back, exposing the stamens. The fruit is a five-cornered berry, giving this beautiful shrub its common name.

For me, this is the most pivotal of all the remedies, for it addresses those universal lessons of self-esteem and self-acceptance. This remedy concerns the celebration of one's physical being and essence. It helps us feel confident about our inner and outer beauty.

During workshops I often ask people whether or not they have ever had a period in their lives when they had low self-esteem or a lack of confidence. The only person who has ever said no later confessed to being a compulsive liar!

Five Corners can be seen as a five-pointed star representing a person in the anatomical position, with the head, arms and legs extended outwards. With positive expression, the energy flows back and forth between those five points, recharging and vitalising the person. When there is a lack of self-esteem the energy flow is blocked, with the five-pointed figure resembling a person "scrunched up", their legs and arms tightly tangled.

People who don't feel good about themselves often give the impression that they are trying to look inconspicuous. This can be reflected in the colours of their clothes, with bland, neutral tones predominant. Those colours blend into the background so that these people don't stand out and aren't noticed. Not only do they dress as drably as possible, but their clothes often cover so much of them that their physical selves are simply negated.

In colour therapy, red stimulates the circulation, and by wearing red you are noticed by others. Low self-esteem is closely correlated with low blood pressure, so by wearing red you can boost your blood pressure and your self-confidence.

Five Corners helps to bring about, first, an acceptance of the self, and then an appreciation of the beauty of the self on all levels.

I had found Five Corners in flower and was drawn to it, sensing very strong properties. A few weeks later at a workshop I was doing some one-to-one work with a woman when she suddenly began to sob violently. She had just realised how much she had suppressed her femininity over the years.

She had grown up very close to her father, but their relationship was based on her acting like a tomboy. Her clothing was masculine, and her friends had never seen her in a dress. She had never celebrated her own beauty.

When she broke down, I held her forehead on the emotional stress release points and, while I had my hand there, a picture of this flower came into my mind and stayed. I knew that this was a remedy for self-esteem and an appreciation of our own beauty.

The woman was quite excited when I described the plant to her. After taking it, she had some sessions on healing clothing with Kristin. They went through her wardrobe, an activity Kristin refers to as the Glad Bag and tissue technique. Any clothes that don't make the person feel beautiful are thrown out, even those that have hung around just waiting to come back into fashion. The tissues are for the tears that come from releasing old possessions.

Then, much to her friends' surprise, she had some dresses made and got a new hairstyle. As she started to appreciate herself she attracted a relationship—something she had always had great difficulty in establishing—with a man who really cared for her. You can't let in love from others if you don't love yourself.

Have you ever been in a relationship with someone who doesn't feel good about themselves? They don't love themselves and they can't accept your love either. This is a very common pattern.

Five Corners increases a person's self love and vitality. As one woman said, "I would love to have an intravenous drip of this essence; I feel so good on it." And with this remedy you can actually see the changes occurring. When people feel good about themselves, they bloom, and it shows. No other essence has the capacity for releasing as many negative beliefs.

I am now celebrating my unique strengths and beauty.
I am becoming more loving and accepting of myself.

Negative condition

•

low self-esteem, especially concerning the physical body

•

dislike of self

•

crushed, "held in" personality

•

clothing drab and colourless

Positive outcome

•

love and acceptance of self

•

celebration of own beauty

•

joyousness

FLANNEL FLOWER

(Actinotus helianthi)

The fastest way to freedom is to feel your feelings.—*Gita Bellin,* Amazing Grace

Actinotus helianthi grows in rocky and sandy areas along the New South Wales coast and up into central Queensland. The entire plant—flowers, buds, stems and foliage—is covered with a soft, silky down and feels like flannel, hence its common name. Its generic name *Actinotus* means "with rays from the centre", referring to the form of the "petals", which are velvety white and tipped with sage-green. These are in fact bracts and enclose a cluster of tiny flowers which form the plant's grey-white centre. The beautiful grey-green, divided leaves are velvety too. The plant flowers predominantly in spring and summer.

Flannel flowers belong to the large Apiaceae family which includes carrots and parsley. With the exception of one species from New Zealand, flannel flowers are found only in Australia.

The soft, sensuous texture of this plant, which begs to be touched, gives an indication of how it works with people,

as Flannel Flower is for people who don't like physical contact.

Flannel Flower is for those who feel uncomfortable about touching or physical closeness. All people have a sense of personal space, some needing a larger area around them than others. However, those people who don't like being touched often crave closeness. They don't get close to others easily, but when they do they are unable to deal with it. They feel as if their space has been invaded. These people don't like being hugged or touched, so social contact or being in crowded places can be very uncomfortable for them.

Flannel Flower helps people enjoy physical contact and activity. It is a remedy for physical expression, either through touching or through movement like dancing or sport. The lazy cartoon character Norm, who stars in the "Life. Be in It" advertising campaign, would be a suitable candidate for Flannel Flower. Norm is overweight and sits watching television with a can of beer in his hand while his wife is in her leotard, jogging on the spot next to him and saying, "Come on, Norm, let's go for a jog." Norm's response is: "Oh, maybe a bit later."

Flannel Flower increases physical energy. It also helps people express their feelings verbally and develop sensitivity and gentleness in touching, whether in a sexual or sensual way.

The Australian male is often portrayed as very rough and insensitive. Flannel Flower allows this type of man to become more comfortable in expressing the gentler, softer side of his nature. It helps him get close to others, whether they are friends or lovers, and share his feelings with them. In many ways Flannel Flower tends to be a male remedy, yet it will work equally as well for women with these patterns.

One area in which Flannel Flower definitely has an affinity for men is that of sexual trauma or abuse. Current figures indicate that 30 per cent of men are sexually abused at some time in their lives, whether this is rape or another form of sexual abuse.

Flannel Flower can help these men feel comfortable about being touched so that they can experience physical contact again. It can even be used by men who have been physically assaulted. I know from personal experience that, after such a trauma, you feel very uneasy with other people. You don't trust others' motives, and for some time you may feel very vulnerable and exposed when close to others. In these situations, Flannel Flower combined with Fringed Violet gives quick, long-lasting results. (For women who have been sexually abused, Wisteria and Fringed Violet work very well together.)

I now enjoy expressing and sharing my feelings.
I am open to physical gentleness and sensitivity now.

Negative condition

- dislike of being touched

- agoraphobia

- lack of sensitivity in males

Positive outcome

- gentleness and sensitivity in touching

- openness

- expression of feelings

- trust

- joy in physical activity

FRINGED VIOLET

(Thysanotus tuberosus)

*B*e at peace and see a clear pattern and plan running through all
your lives,
Nothing is by chance.
— *Eileen Caddy,* Footprints on the Path

This plant is one of twenty different *Thysanotus* species found
in every State of Australia except Tasmania. In the southern
States it is referred to as Common Fringed Lily, while in New
South Wales and Queensland it is known as Fringed Violet.
It belongs to the vast lily family, though Australia doesn't
have any lilies in the true botanical sense.

Fringed Violet is an erect, slender plant up to 14 centimetres
high, with grasslike leaves. Long, smooth stems carry the
beautiful mauve to purple flowers which consist of three
petals, each about 7 millimetres long and delicately fringed
with fine cilia. The plant blooms in spring and summer. Fringed
Violet prefers open, sunny areas, such as the grassy heathlands
of sandstone country. Its beauty is transient, as the elegant
purple flowers open for only a single morning.

I can still remember the joy of finding this plant for the first time. I had often flicked through the pages of books to photos of this flower and had always been impressed by its beauty, but it was certainly no disappointment when I found it in the heathlands.

Like so many of the other plants, the Doctrine of Signatures is clearly shown in Fringed Violet. The hair-like cilia resemble an aura, which indicates one of the plant's main functions: to help restore a person's aura after it has been damaged by shock or trauma. The shock may have been caused by the loss of a loved one, bad news or an unexpected event. Fringed Violet helps people deal with shock so that it does not affect their inner peace. It prevents them from being thrown off balance by external events.

With a little practice, it is quite easy to detect a break in someone's aura. Just run your hands around and slightly above the person's body. Any cold spots indicate that there are breaks or weaknesses in the aura at those points.

The aura can also be affected by electromagnetic radiation from power lines and transmission towers. It has been shown that the cells in a healthy body rotate clockwise, but under these negative influences the direction can be reversed, indicating a poor state of health and vitality. Fringed Violet combined with Crowea, Waratah, Paw Paw and Bush Fuchsia is excellent for negating the effects of electromagnetic radiation.

Fringed Violet combines well with other remedies for certain purposes. For physical or sexual assault it is used with Flannel Flower for males and with Wisteria for females. It can also be used with Grey Spider Flower for psychic protection, as when the aura is broken we lose our protection and negative energies may enter us. As we approach the end of this century the negative influences around us are increasing, so this remedy well become more and more important to us, as will Waratah essence.

For some time after birth a baby's aura is open, like its anterior fontanelle. Putting a few drops of the Fringed Violet essence on this spot and making the sign of the cross over it will help to close the aura to negative influences. Some midwives advise that a child should not be taken out of the house for six weeks after birth for that reason. This precaution is unnecessary after using Fringed Violet, as the protection is immediate. It can be applied once a day for a few days.

Fringed Violet is very useful for someone who has never felt well since a particular event such as surgery or bad news, as it is able to release an individual from a long-standing trauma. After an amputation, some people experience a curious phenomenon called phantom limb pains in which they have sensations apparently from the part of the body that was amputated. This can be explained by Kirlian

Negative condition

- damage to aura

- shock, trauma

- lack of psychic protection

- poor recuperation since trauma or shock

- fear of physical contact since rape or assault

Positive outcome

- removal of effects of recent or old trauma

- reintegration of physical and etheric bodies

- psychic protection

photography. With this technique, if a leaf is cut in half, a photograph of one half will show an aura around the whole plant, including the half that is no longer there.

We have all heard the expression "to jump out of one's skin". This is literally what happens with fright—the inner and outer bodies become unbalanced. Fringed Violet helps to realign these bodies after a shock or trauma.

People who are in shock recuperate much faster if Fringed Violet is taken. It is advisable to give either Fringed Violet or Emergency Essence, in which Fringed Violet is one of the ingredients, to anyone who has been involved in an accident, as even if the person doesn't appear to be in shock, there may be a delayed reaction. Sometimes even the shock of bad news can be stored in the body for weeks afterwards, and it is not uncommon for people to develop eczema or skin rashes some time later.

When a person has a major illness, the cause may be found in a deep negative emotion or trauma eighteen months earlier, as sometimes it takes this long to manifest physically. Starting in the outer bodies and working its way inwards, cancer, for instance, follows this pattern. Unresolved shock can also manifest in a nervous breakdown. Fringed Violet goes back to the time of the shock and helps to neutralise it.

There may be many indications for Fringed Violet in the eyes: unequal pupil size, very large pupils, the nerve wreath shattered and also a lack of clarity in the border between the iris and the white of the eye, indicating a person who is too easily influenced by others or external events. Fringed Violet helps to keep intact a person's protection, thus blocking off unwanted external energies. It is excellent for people who are drained by others, or those who unconsciously absorb the physical and emotional imbalances of other people.

In age recession and especially past-life regression work, Fringed Violet is very useful for bringing a person back to the present time at the completion of a session. It is very good, too, after a rebirthing session or after a number of cathartic or deep emotional experiences have been shifted. It is also a good remedy for birth trauma, as birth is usually a time when individuals are at their most susceptible and sensitive.

My life force now radiates good health and vitality.
I am balanced and integrated and have universal protection now.

GREY SPIDER FLOWER

(Grevillea buxifolia)

Come to the edge, he said.
They said: we are afraid.
Come to the edge, he said.
* They came.*
He pushed them . . . and they flew.
          —*Guillaume Apollinaire*

There are over 250 species of *Grevillea* found across Australia, ranging from small ground-hugging varieties to tall trees like the Silky Oak. They all belong to the large Proteaceae family. *Grevillea* is named after C. Greville, an English botanist, while *buxifolia* refers to the shape of the leaves which resemble those of the European Box (*Buxus*).

Grey Spider Flower is a common shrub, 1 to 2 metres high, which is found around the Sydney and Blue Mountains areas. The rusty brown flower head is covered with a fuzz of greyish white hairs. It is made up of masses of individual florets crowded into rounded clusters at the ends of the branches. Altogether, it looks like a rather untidy spider. The long curved parts, like the spider's legs, are the pistils of each flower. The flowers are rich in edible nectar, and the shrub blooms for most of the year.

When you look at this flower you can clearly see a face, with two sunken eyes and a wide-open mouth, resembling the famous, haunting, expressionistic painting of the 1930s by Edward Munch, *The Scream*. This suggests the quality of the plant, as it is very good for helping to resolve terror and to bring about courage, calmness and faith.

When making the essence, every time I went to pick a flower I would find a green spider in it. According to Rudolph Steiner, the group-soul aura of the spider family is so striking that very few people with clairvoyance can look at it. It seems that some aspects of nature are just too different for us to be able to appreciate their great beauty, hence our fear. From a psychological viewpoint, there is also the theory that spiders are an archetypal symbol of primordial fear. In my childhood, my most terrifying experience was watching huge spiders in a tropical jungle in an adventure film. We also feel afraid because of the actual physical danger to us from poisonous spiders. Certainly anyone who has walked into a spider web can testify to the initial feeling of fear.

Negative condition

•

terror

•

fear of supernatural and
of psychic attack

Positive outcome

•

faith

•

calmness

•

courage

The Grey Spider Flower remedy is for extreme terror, which is much more intense than the minor, insidious fears associated with Dog Rose. This is absolute and immobilising terror or panic. It may be the fear of dying or of losing our identity, or the fear of things so shocking, so terrifying, that we may not survive them.

Some people's greatest fear in life is public speaking. They become extremely nervous and often shake violently or are unable to speak in front of a group of people. This is different from the fear associated with Dog Rose, which is a much more general fear.

Terror directly affects the physical body in a very obvious way. The person has wide-open pupils, a dry throat and a pounding heart. The adrenalin reaction to a terrifying situation is what leaves a person feeling drained afterwards. In the iris, the zone for movement coordination is next to the anxiety area, and when there is an overload of terror or anxiety it will spill into the next zone, the movement zone, so the person won't be able to move, being paralysed with fear.

One man whom I treated was terrified of being alone in enclosed spaces. He was working for the gas company and often had to go into all sorts of places—cellars, for instance— so his fear was obviously of the Grey Spider Flower type.

This essence can be used when there is a great deal of terror in the atmosphere, such as in wartime. A technique that can help when you are watching reports of horrifying events or gruesome accidents, or, even worse, if you are involved in one, is called Buttoning. If you place your tongue on the roof of your mouth behind your front teeth, these events won't drain your energy level as much.

I remember when my uncle saw a horrible accident in which three people were killed. He came home shaking with shock and terror, very much aware of his own mortality. Grey Spider Flower with Fringed Violet would have been excellent for him at that time.

As already mentioned, Grey Spider Flower can be used with Fringed Violet for protection against psychic attacks, or at times when you feel your life is threatened. These remedies will help to develop trust and faith and the knowledge that the Light can always be called on to protect you.

As with Dog Rose, Grey Spider Flower can be used for nightmares, especially in young children whose fear often stays with them long after they have woken up. Many children experience terror as a result of nightmares, while adults are not so disturbed by them. Quite often adults underestimate the effect of nightmares on children.

Recently I gave this remedy to a five-year-old child who became terrified and physically immobilised if he was outside when it rained, and Grey Spider Flower quickly removed this fear.

Grey Spider Flower is for terror that comes on suddenly as a reaction to something horrible, and it works quickly to calm people and help them recover. It can be used for severe asthma, to help the person cope with the fear that they are going to suffocate, and for severe phobias. For milder forms of these disorders, Dog Rose can be taken.

The Grey Spider Flower remedy helps to bring about faith, calmness and courage. These qualities can be seen in people who work in very dangerous situations such as bomb disposal and police rescue units where great courage is needed for these people to put their own safety second to the welfare of others.

Divine power protects me now.
Faith, calmness and courage are now mine.

HIBBERTIA

(Hibbertia pendunculata)

This is your life and nobody is going to teach you,
no book, no guru.
Learn from yourself, not from books.

It is an endless thing and when you learn about yourself from yourself,
out of that learning wisdom comes.
—Krishnamurti in A Bag of Jewels
(eds Susan Hayward & Malcolm Cohan)

The hibbertias include over 115 species in a genus that, with few exceptions, is confined to Australia. All the species of this genus have yellow flowers which bear a strong resemblance to one another. This has led to their common name of Guinea Flower, as the flowers "gleam like golden guineas".

Hibbertia pendunculata is a profuse, low, trailing shrub. It is widely found in open forests and cleared areas in the mountainous and coastal areas of New South Wales and eastern Victoria. Its flowers are 12 millimetres across, grow on long stalks and bloom in spring. The five petals, when they fall, leave a heart-shaped form on the ground. The shining, dark green leaves form a strong contrast against the bright yellow flowers.

Like most yellow-coloured flowers, this remedy deals with the intellect. Hibbertia is for people who devour information and philosophies. These people have a great desire to learn, so they are constantly reading books and doing courses. They are very strict with themselves, especially in their pursuit of knowledge, even to the extent that they become fanatical.

Yet the purpose of all this striving for knowledge is to improve themselves so that they can feel superior to others. Although they may seem to integrate all the ideas and information they absorb, it becomes purely "head stuff"— an intellectual exercise. They are not looking for a practical way to use this information but believe they will be better

people because they have more knowledge than others. They gather information from here, there and everywhere, grabbing a bit of the philosophies of many different teachers and gurus. As this information doesn't come from their own experiences, the end result is that they feel weighed down by it, without any real understanding or knowledge.

As Albert Einstein said, "Logical thinking cannot yield us any knowledge of the empirical world; all knowledge of reality starts from experience and ends in it."

The energy of these people is focused in their heads more than in any other part of their bodies, especially their hearts. The men in this category are often quite tall and thin and have a tendency to lose their hair at the front. In some of these people, their noses will actually grow longer.

The positive aspect of the Hibbertia essence is the integration of information and ideas with one's experiences and intuition, in order to achieve a balance. As a result, one accepts and trusts oneself as well as the knowledge that one has and uses, without wanting to feel superior to others.

Hibbertia is for people who are inflexible and dogmatic, especially in adhering to their ideals. They also like to be in control, exhibiting a lot of self-denial and self-repression. This rigidity of the mind is often reflected in a stiffness of the body, both of which are manifestations of a lack of flexibility and hardened attitudes.

One night, during my seven or eight years of being a strict vegetarian, I was at a dinner party. Among the vegetables I found tiny pieces of fish which I diligently and pendantically picked out before I ate my meal. In hindsight, I think I was being quite fanatical.

As the Hibbertia flower fades, its five yellow petals drop to the ground and leave heart-shaped patterns imprinted in the mud. It is as if the flower is making a final plea to us to remember the importance of the heart in our lives. In fact, one of the positive outcomes of this essence is reconnecting the heart and the head. As Gita Bellin writes in *Amazing Grace* (Book One), "When the heart feels and the mind understands—then we become a whole human being."

Numerologically, the number 5—the number of petals on the Hibbertia flower—represents emotional centring, and the Hibbertia essence helps to drain excess energy from the mental plane in order to balance the emotional plane.

As the Hibbertia essence can teach us, education is not parroting texts but is revealing truth from the highest inner sources.

I now accept and integrate my inner wisdom.
I am flexible in all life situations now.

Negative condition

•

self-improvement
fanaticism

•

addiction to acquiring
knowledge

•

excessive self-discipline

•

feeling of superiority

Positive outcome

•

content with own
knowledge

•

acceptance, ownership
and utilisation of own
knowledge

ILLAWARRA FLAME TREE

(Brachychiton acerifolius)

*You will never be judged unless you accept the judgement
of those around you.
And if you accept their judgement, it is only your will to do
so for the experience.*
— *Ramtha,* Ramtha

This tall, spectacular tree occurs in coastal rainforests from
Illawarra south of Sydney to Queensland. In its natural habitat
it can grow as high as 30 metres. A mature Flame Tree has
thick, wrinkled, grey outer bark and lacelike inner bark.

As the tree comes into flower, the leaves drop to expose
a myriad of waxy, scarlet, bell-shaped flowers hanging in large
clusters on similarly coloured, branching stems. The whole
tree blazes as if aflame when in full bloom in early spring.
Then new green leaves, up to 25 centimetres long, develop
as the flowers wither. Older trees bear oval leaves, while the
young trees have leaves divided into five lobes which look
like a hand reaching out.

Because of the properties of this essence, it wouldn't be
surprising if the outstretched hand was asking to be accepted.
The Flame Tree essence is for people who often feel rejected.
They feel left out quite easily and are very deeply hurt when
they do perceive rejection—real or imaginary. When others

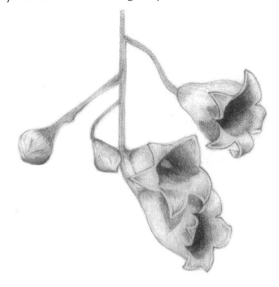

would think, "Thank goodness I have a bit of time to myself," these people feel that no one loves them if they are not included. Consequently, they often do things they would rather not, to avoid possible rejection. Such actions are a deep denial of self and lead to a weakening of the thymus gland, the key to the immune system. Flame Tree essence strengthens and balances the thymus. They also have a tendency to reject themselves, which can lead to a state of dejection.

It has been said of Flame Tree people that if they were the sole survivors of a cataclysm, they would feel tremendously rejected because everyone got taken but them.

These people know they have certain abilities but never get around to developing or using them. They ignore their potential because they feel overwhelmed by the responsibility of developing it. They put it off until tomorrow—and, of course, tomorrow never comes.

The Flame Tree remedy will help these people take the first step towards realising their potential. It will allow them to make a commitment to a certain course of action, and they will find the confidence and strength to deal with what they need to do in life, without being overwhelmed by the responsibility of it.

One man who felt a great sense of rejection took this remedy with excellent results. He had been reluctant to try anything new. Now he is much stronger in identity and feels he can make any move he wants to without feeling rejected.

Red Grevillea people know what they want to do but don't know how to achieve it. Flame Tree people know what they need to do but feel overwhelmed by the responsibility of doing it.

These people also continually delay parenthood. They want children but are afraid of the responsibility. An easy way to check whether or not this is the case is to ask them how they would feel about having twins or triplets, and then watch their body language. They may say they could cope but will begin to look very restless and uncomfortable.

Many of the people for whom I have prescribed this remedy have grown up with a Flame Tree either in the backyard or flowering close by.

Flame Tree is a good remedy for children at school. It will help them deal with temporary setbacks such as missing out on the cricket team, or being excluded from the "in" crowd.

It may be useful for those children who have started at a new school where the teachers and other children don't pay much attention to them. The type of child who could use the remedy will perceive this situation as rejection, and rather than trying to make friends, will become despondent.

I am loved and accepted in all life situations now.
I now accept responsibility gladly.

Negative condition

•

fear of responsibility

•

overwhelming sense of rejection

Positive outcome

•

confidence

•

commitment

•

strength

•

self-reliance

•

self-approval

— 90 —

ISOPOGON

(Isopogon anethifolius)

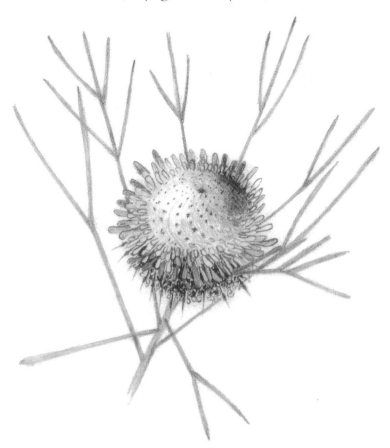

We learn wisdom from failure much more than from success.—
Samuel Smiles

Isopogons belong to the diverse Proteaceae family. There are
approximately thirty species in the genus, the majority being
found in Western Australia. They grow in all States except
Tasmania, in open sclerophyll forests and heathlands.

In this genus the flowers are crowded into cylindrical
clusters around a woolly stem and a domelike seed head.
The common name Drumstick alludes to this seed cone on
the end of the stem. *Isopogon anethifolius* is an erect shrub up
to 3 metres high, with yellow flowers 12 to 25 millimetres
in diameter. As the flowers wither, they reveal round grey
seed cones which stay on the plant for many years. This

shrub has finer, narrower foliage than the other species of the genus, with rigid leaves 4 to 6 centimetres long.

In metaphysics, yellow symbolises wisdom and knowledge, and Isopogon, like the yellow-flowered Hibbertia, deals with the intellect. In fact, Isopogon has many properties that involve the mind. It is very useful for opening up the subconscious mind and allowing the retrieval of long-forgotten skills.

Waratah has similar aspects, but it is used in situations of crisis or emergency when old *survival* skills are desperately needed, whereas the Isopogon remedy is more appropriate when a person wants to remember a skill or information learnt years ago, such as speaking a foreign language.

Isopogon will help to retrieve memories of the past. My work with age regression has proved to me that all our experiences are stored in the mind. We can remember any event by using the right key to unlock the door to the subconscious. This remedy has also helped those suffering from senility, Alzheimer's disease and poor memory, but it may need to be taken on and off for a few months.

Isopogon can also be taken by people who fail to learn from their experiences. Nothing happens to us by chance. If we don't learn from one event, then the lesson will be repeated over and over again until we finally understand it.

Those people who don't learn from their mistakes often race through life trying to dominate and control it, without stopping to review previous experiences. One of the best ways to learn is to make many mistakes, but after making each one, to correct our behaviour accordingly.

The space flight that took Armstrong to the moon was on course only 3 per cent of the time. It managed to land on the moon because its position was constantly monitored and corrected to keep it on course. That is what life is, too— constantly reviewing our actions and their consequences and correcting our behaviour according to what we have learned.

A case history illustrates this point: "I took this essence because I had the feeling that my head and my heart were not working together and I did not remember whether I was learning from my experiences. Three months later I had a big breakthrough. Suddenly I could understand things like drawings that I did, dreams that I had years before. I felt like a thick curtain had been peeled away from my eyes. Since then I am very consciously aware that everyday life is a learning process and only sometimes I have the feeling that head and heart are separated."

This remedy is also useful for people who need to dominate and control others. They want to be in charge at all times and can't imagine that others may be capable of doing a job as well as or better than they can. They often believe that they know more than others, or that they know better. Their characteristic expression may be: "Don't do as I do, do as

Negative condition

- poor memory
- inability to learn from past experiences
- senility
- controlling personality
- manipulative personality

Positive outcome

- ability to learn from past experiences
- retrieval of forgotten skills
- relating, without manipulation or control

I say." This can be seen in very rigid parents. They may have a very arrogant, bossy, domineering attitude towards others.

These people are often dominated by their intellect and cut off from their feelings. They may be very powerful, ambitious, demanding and tyrannical and intolerant towards weaker people. Moreover, they may be very stubborn for long periods of time in order to control others.

The positive aspect of this essence is the natural leader who is inspiring, wise, tolerant and flexible and who encourages others to develop their own skills and abilities. This is a person who learns from experience.

I now learn from my experiences.
I now relate to others with understanding and flexibility.
I now have access to my skills whenever they are needed.

JACARANDA

(Jacaranda mimosaefolia)

The shortest way to do many things is to do only one thing at once.— Samuel Smiles

The Jacaranda is a native of both Brazil, where it likes the dry, high areas, and the West Indies. It is not only the colourful blossoms that make this tree so attractive, but also the foliage, with its delicate, fernlike, lacy appearance. This tree has been widely cultivated on the east coast of Australia where its lavender-blue blooms light up the landscape and streets in late spring.

Jacaranda trees grow up to 30 metres high and are deciduous. The loose clusters of numerous bell-shaped flowers are about 20 centimetres long and drop off to form a mauve carpet around the trees. The light green terminal leaves are 30 centimetres long and grow in opposite pairs. Each leaf has sixteen or more pairs of leaflets, and each of these has about twenty pairs of tiny leaflets.

Jacaranda is a remedy for ditherers, those who are always starting projects but rarely finish them, mainly because they are so easily distracted. They have great difficulty in making decisions, as they constantly change their minds. There is a very scattered feeling about the Jacaranda type. They tend to be all over the place, rushing about here and there. Interestingly enough, the town on the north coast of New

South Wales that hosts the Jacaranda Festival, Grafton, is in an area which the Aboriginals considered to be an undesirable place to live because its energy was too scattered. The profusion of Jacaranda trees now growing there may have helped to focus that scattered energy.

Jacaranda is similar to Black-eyed Susan, as both types of people undertake many different projects. But the Black-eyed Susan type is full of energy and focuses on getting things done, quite often more than one project at the same time, whereas the Jacaranda type tackles just as many projects but lacks focus. The projects remain in various stages of incompletion. The home of this person may have a half-painted room or a partly built set of shelves. Yet this type is enthusiastic, inspired and filled with excitement at the idea of a project but just doesn't follow through.

It is very draining to be around Jacaranda people because they are so scattered. They often have many acquaintances who enter and then leave their lives. Others tend to get caught up in their initial enthusiasm, yet these Jacaranda people may suddenly change their social circle or their acquaintances may see through them and quickly drop them.

Such people often move from house to house. They are insecure about making the right decisions, which leads them to vacillate. At other times they panic, thinking they have made the wrong choice, and so change their minds.

They change their health practitioners or doctors regularly and often don't complete a course of treatment. In the irises

Negative condition

- scattered

- changeable

- dithering

- rushing

Positive outcome

- decisive

- clear-minded

- quick-thinking

- centred

of Jacaranda types the fibres, instead of being straight, are often crunched up, indicating that their energy is not flowing freely. They do not flow with life but zigzag through it. Their illnesses are those in which the symptoms change a lot. They may have pains that move about or suffer from nervous rashes or hives. They tend to be accident prone, as being unfocused they can be quite clumsy. This aspect of Jacaranda is similar to Sundew. But the Sundew remedy is for people who are very vague and dreamy, not for those who rush around.

The positive aspects of the Jacaranda essence is a person who is poised, decisive, clear-headed and quick-thinking. These people are prepared to consult others and listen to advice, but then make up their own minds in view of all the facts of the situation. They are flexible and can judge whether or not a certain course of action is likely to achieve the desired results. They can put strategies into action and their projects are often successfully completed.

I make decisions quickly and easily.
I am now clear-headed and quick-thinking.

KANGAROO PAW

(Anigozanthos manglesii)

Ignorance is not bliss—it is oblivion.—Philip Wylie, Generation of Vipers

There are ten species in the *Anigozanthos* genus, which was first described in 1800 by French botanists. They all occur in the south-western corner of Western Australia. The original description of the flower was of "a corolla in the form of a tube divided in the extremities into six unequal parts".

The species epithet *manglesii* honours Robert Mangles, an English horticulturalist who successfully grew the type specimen from seed brought to England in 1833 by Western Australia's first governor, Sir James Stirling.

The green flower bud of the green and red Kangaroo Paw opens and tips fold back to reveal six golden stamens and a style stretching even further forward. Deep in the flower nectar is found. Each flower can be seen to represent the front foot of a kangaroo, even to the fur that encloses the petals.

Kangaroo Paw is a fairly short-lived herbaceous perennial. The flowering stem can be up to a metre long and blooms from August to October. The plant occurs naturally from the Murchison River around Kalbarri to Manjimup in the south. It is commonly found in the strikingly contrasting mauve-red and green form.

Kangaroo Paw thrives after bushfires or land clearing. Over a period of time other long-lived plants crowd out Kangaroo Paw, which will not be seen again until the next fire or clearing.

In 1960 it was chosen from over 8000 plant species native to Western Australia, to be the State floral emblem. Such a choice would not have surprised the late Joseph Furphy who, fifty-five years earlier under the pen-name Tom Collins, described the flower in his book *Such is Life* as follows:

This is the flower which is so much like Australia. None of your Mountain Daisy business about it, mind; no

Primrose or Snow-drop racket; but a splendid, confident, audacious glory, leaving even the black and scarlet Desert Pea a bad second. Even the unpoetical Gropers claim the Kangaroo Paw as their national flower.

This is another example of a State floral emblem that was first chosen and only later officially sanctioned.

My trip to Western Australia was memorable, not only for the most spectacular wildflowers in the world and the beauty of the land, but also for the friendliness and hospitality shown to me by so many people. The Kangaroo Paw remedy seems to encapsulate those qualities. It is for sensitivity to the needs of other people—for kindness. It also provides the ability to enjoy and feel comfortable with people with different backgrounds and personalities, and to understand what they need so that they, too, can feel relaxed and comfortable. An example is a good host, who makes one feel very much at ease, or a proficient guide, who familiarises people with local customs.

The negative aspect of Kangaroo Paw is someone who is very green—after all, the flower is partly green! These people are very clumsy socially and have great difficulty in relating to or dealing with others, because of naivety or narrow-mindedness or because they simply feel out of place. This remedy has been used very successfully with insensitive people, who put excessive demands on others or who are blind to or consciously ignore the needs of those around them. These people certainly don't know how to make others feel comfortable. Consequently, they create great tension in social situations. A common response from those who have taken Kangaroo Paw is relief. They have worked through their awkwardness and can now enjoy other people. They feel so much lighter and relaxed.

Kangaroo Paw types may bear the brunt of ridicule in social situations. It is not unusual for them to be unaware of others laughing at them or putting them down as they are so self-absorbed. Or they may realise that they do not cope well in company and may avoid contact with others. They may also become defensive when they are around those with whom they don't feel comfortable and who don't feel comfortable with them.

Kangaroo Paw can also be taken by the person who always knows the right thing to say after the opportunity has passed. It is a remedy that helps people understand how to act appropriately in all situations.

I now relate with poise and sensitivity to other people.
I now feel comfortable with other people and enjoy their company.

Negative condition

- gauche
- unaware
- insensitive
- inept
- clumsy

Positive outcome

- kindness
- sensitivity
- savoir-faire
- enjoyment of people
- relaxed

KAPOK BUSH

(Cochlospermum fraseri)

*N*ever give in. Never give in. Never give in.—*Winston Churchill*

This small, slender tree usually grows up to 3 metres high and is found on rocky ground, such as hillsides and ridges, in the tropical parts of Queensland, the Northern Territory and Western Australia. Its common name relates to the large, fragile seed pods which when ripe are filled with a dense, soft, cottonlike material. The Aborigines used this material for body decoration. Later settlers used it for stuffing saddlebags and pillows.

The generic name *Cochlospermum* is derived from the Greek words *kocklo*, meaning wind, and *sperma*, meaning seed. This refers to the little "parachute" attached to each seed, which aids its dispersal by the wind.

The Kapok Bush is a common sight along roadsides during the dry season when its loses its leaves and its large yellow flowers are prominent. The flowers have five petals, hairy sepals and many stamens and bloom from May to September. They are very aromatic and are edible.

The Kapok Bush essence was made up at Mount Barker in the Kimberleys on an Aboriginal cattle property. It is a remedy for commitment and following through, for never giving up and not accepting defeat. Its positive aspect can be seen by someone who masters a technical problem or works out how a complex piece of machinery operates. These people have an overview of the problem and work through it sequentially to a solution.

Even in the dry, hot air of the Kimberleys in the third year of a drought, the Kapok Bush could be seen still flowering. The flowers have a delightful, slightly mucilaginous taste. On many of my bushwalks eating the flowers kept me going.

This remedy is for people who have a tendency to give up very easily and not follow things through. When effort is required they become discouraged or despondent and abandon the task. With Kapok these people are much more willing to give things a go.

The yellow of the flower hints at its quality of intellectual understanding—to be able to persevere in order to logically solve a problem. For example, if something is broken, rather than just giving up and abandoning it, this remedy gives one the insight to analyse its workings and possibly come up with a way to repair it.

The negative aspect of Kapok Bush is a lack of effort, because everything is too much trouble. This type of person tends to have a dampening effect on joyous or exciting atmospheres. But what you put out you attract, so this person attracts people with similar natures, which can make life pretty dull.

The attitude of giving up is reflected in certain illnesses, such as the wasting diseases of anaemia, cancer and low blood pressure, a constantly exhausted feeling or immune system deficiencies. When people are told that they have cancer, they either fight it or else they die very quickly because they just give in and become resigned to their fate. When people with no fight in them contract a minor illness, their recovery is slow.

The best way to overcome an illness is to desperately want to be well and able to do all those things that the illness is preventing you from doing.

This essence has worked wonderfully well not only for illnesses, but also in the classroom situation where parents and teachers have given it to children who don't try if they think a subject such as maths is too hard.

The positive qualities of this remedy are evident in the Aboriginal community at Mount Barker, which is teaching its people mechanical, managerial and other skills from which the community is making a very good profit. These people think problems through and repair their own machinery. They are not overawed by new technology but rather make it work for them.

Negative condition

·

apathetic

·

resigned

·

discouraged

·

half-hearted

Positive outcome

·

willingness

·

application

·

gives it a go

·

persistence

·

perception

When people work in jobs they don't really like or suffer from long illnesses, their life force may become so weak that they lose their fighting spirit. They just won't, or feel they can't, take control of their own lives. They tend to be apathetic or half-hearted about tackling life's problems. As we create everything that happens to us, we are, in fact, able to change our lives for the better.

The Kapok remedy helps people respond positively to the challenges of life. It allows those who lack commonsense to become more practical and down-to-earth and helps them apply themselves to problems and persist until they find solutions.

When we were at Mount Barker we visited a magnificent waterhole. It took quite a number of hours to get there. One night we asked whether anyone had been there, and a number of people replied, "Oh, no, it's a bit far to walk, isn't it."

In the whole of the Kimberleys we had not seen a more beautiful spot. So I suppose the reward for our persistence was to have the enjoyment of that waterhole all to ourselves.

I am now willing to respond positively to life's challenges.
I now take control of my own life.

LITTLE FLANNEL FLOWER

(Actinotus minor)

If I had my life to live over, I would start barefoot earlier in the spring and stay that way later in the fall. I would go to more dances. I would ride more merry-go-rounds. I would pick more daisies.—Nadine Stair, "If I Had My Life to Live Over" in Chop Wood, Carry Water *(ed. New Age Journal)*

This plant is a much smaller version of the Flannel Flower (*Actinotus helianthi*), but its flowers have a similar velvety texture. Little Flannel Flower has dense, woolly white flowers surrounded by spreading bracts, which are clustered on small, wiry stems and are less than 12 millimetres in diameter, a sixth of the size of *A. helianthi*. This delicate, spreading, perennial herb blooms most of the year.

Little Flannel Flower is commonly found in heathlands and open forests on rocky hillsides, from the Hawkesbury sandstone region to the Blue Mountains in New South Wales. However, as it is often hidden by larger plants, it can be easily overlooked. A field of these flowers produces a sense of tremendous lightness.

This remedy addresses the child within us all, whether we are adults or still children. It produces an expression of playfulness and carefree, spontaneous joy. It can be given to children who are growing up much too quickly and who tend to take on the troubles of the world. It is also appropriate for children who watch a lot of television or listen to the news, since much of what they see or hear concerns negative events like earthquakes, starvation and war, or dramas involving violence and/or unhappiness. Their picture of the world is coloured by what the media presents to them, so that they perceive it as a very sombre place full of grief, destruction and negativity.

Many of these children have become old before their time, missing the opportunity to enjoy their childhood. Little Flannel Flower helps them recapture their innocence and childlike playfulness. And, of course, the most effective way for children to learn is through playing games.

Today many courses in business management and leadership incorporate games, as teachers have found that playing is the most effective way of learning. Sometimes playing these games is fun, but they are also very successful teaching aids. It is a great shame that in our culture playfulness

is restricted to the childhood years and is often actively discouraged in adults. Every person has a child within them, and this remedy allows that playful child to emerge.

Little Flannel Flower is a good remedy for adults, especially parents, allowing them to lose some of their inhibitions and to play, enjoy themselves and have fun with children and other adults. Of course, if people are in tune with their gut feelings, they will normally be able to experience the fun and joy of life and will realise that life is not a serious battle. This remedy helps people "trust the dance" and have a lot of fun along the way. Any adult with a rigid mind will end up with a rigid body. Spontaneity and lightness is often lacking in people with physical stiffness, such as arthritis, or heaviness. Little Flannel Flower has allowed many people to feel and look much younger.

As Nadine Stair wrote in "If I Had My Life to Live Over":

> I'd like to make more mistakes next time. I'd relax, I would limber up. I would be sillier than I have been this trip. I would take few things seriously. I would take more chances. I would climb more mountains and swim more rivers. I would eat more icecream and less beans. I would perhaps have more actual troubles, but I'd have fewer imaginary ones.
>
> You see, I am one of these people who live sensibly and sanely hour after hour, day after day. Oh, I've had my moments, and if I had to do it again, I'd have more of them. In fact, I'd try to have nothing else. Just moments, one after the other, instead of living so many years ahead of each day.
>
> I've been one of those persons who never goes anywhere without a thermometer, a hot water bottle, a raincoat, and a parachute. If I had to do it again, I would travel lighter than I have.

This remedy is also very important for connecting small children with their spirit guides and giving them an awareness of the spiritual realm around them. Most children are very psychic and clairvoyant at an early age, yet they learn that it is simply not acceptable to perceive such things. When they do not use their abilities, they eventually lose them. There are many cases of small children telling their parents or friends of a loved one, perhaps a grandmother, who has recently died and appears to these children and speaks to them. Or they may see their spirit guides and talk to and play with them. When these children tell their parents about their experiences, they are often told that they are making up stories or that they are lying. Consequently, the children begin to feel that they are doing something wrong and stop using their gifts.

Negative condition

•

denial of the "child" in the personality

•

seriousness in children

•

grimness in adults

Positive outcome

•

carefreeness

•

playfulness

•

joyfulness

— 102 —

Spirit guides help to guide and direct us. Many of our gut feelings are actually the nurturing of spirit guides. The individual still has to have experiences and learn from them, but life can be much easier when the guide's promptings are heeded. Guides are ready to help us learn our lessons quickly and experience life with as little pain and anguish as possible. If more children maintained contact with their guides, the benefits to our community and our quality of life would be quite staggering. They would also be greatly increased if there were more happy and playful adults able to have fun.

I am now free to express joy and playfulness.
I now flow spontaneously wherever the Universe takes me.

MACROCARPA

(Eucalyptus macrocarpa)

Begin with the possible;
begin with one step.
There is always a limit,
you cannot do more
than you can.
If you try to do too much,
you will do nothing.
—P.S. Ouspensky and G.I. Gurdjieff in
Begin It Now (ed. Susan Hayward)

Locally known as Mottlecah or Rose of the West, this eucalypt grows up to 5.5 metres high. Stands of it naturally occur in a few isolated areas in the wheat-belt of Western Australia, but even these are becoming increasingly rare. Recently, though, it has been successfully cultivated in Europe and California.

This mallee (a shrubby eucalypt usually having several slender stems growingfrom a woody base) was first described in 1842. The specific name it was then given is derived from the Greek words *makros*, meaning large or long, and *karpos*, meaning fruit, for it possesses the broadest fruit, as well as flower, of all the eucalypts. Its solitary flowers, usually deep red, can measure up to 75 millimetres across. They bloom from August to November and make a splendid contrast with the plant's silvery grey leaves.

To see the seed cap flung off by the strength of the red stamens is to experience the properties of this essence. It is a remedy for energy, vitality and physical endurance—the very qualities symbolised by the colour red.

This remedy seems to have an affinity with the adrenal glands—the fight or flight system of the body. When the adrenals are stressed, tiredness and a lack of energy results and the immune system is weakened. There are many causes of low energy, yet on a physical level Macrocarpa can help to recharge and revitalise the body. It combines very well with many of the other remedies, such as Old Man Banksia, *Banksia robur* and Crowea, depending on the actual situation and the individual involved.

In the case of burnout, it could be appropriate for someone with the Black-eyed Susan personality who rushes around far too much and ignores all the telltale signs of stress until, finally, the body collapses. Then, of course, that person's tendency to recreate the pattern would also need to be addressed.

The Macrocarpa remedy is a very good tonic for people who simply need a pick-me-up. When the adrenal glands or one of the other endocrine glands are not functioning well, the whole endocrine system is affected.

A friend of mine at university, who was taking quite a lot of drugs, which didn't do much for his vital force, was working in two jobs at the same time. He had a manic tendency and hadn't slept for about three days when he had a complete breakdown. He lost control and became incontinent while at work. Stress had taken its toll on his

Negative condition

•

convalescent

•

tired

•

exhausted

•

burnt out

•

low immunity

Positive outcome

•

energy

•

vitality

•

endurance

adrenals. This is the type of situation in which Macrocarpa can be of great benefit.

Macrocarpa helps to reinforce the need for rest as well as giving extra energy. It is one of the most widely used remedies in urban environments, and its benefits are experienced exceptionally quickly. A feeling of being more alive, energetic and in control is common after taking the remedy.

Macrocarpa can also be taken at times of great physical stress, when endurance is necessary. It can be alternated with Emergency Essence during childbirth, at exam times and when a great amount of physical energy is needed for activities such as competitive sports and prolonged physical labour.

When people have set themselves a difficult goal, they may be so exhausted after achieving it that they are in no state to enjoy their leisure. In these cases a combination of Silver Princess and Macrocarpa can be very beneficial.

If the body's adrenal glands are exhausted, the immune system is very weak too. The adrenals provide the body's ability to fight back and are very closely linked to the immune system. In the iris, the pupil may be large, with markings in the adrenal and spleen areas. After taking Macrocarpa people may sleep very deeply for a long time, as they are very tired.

It is also a good remedy for people who are convalescing, to strengthen their adrenal glands. It can be combined with half an hour of meditation per day, which is equivalent to two or three hours of sleep, in order to speed up their recovery.

I am now experiencing a renewal of energy and vitality.
I am now tapping into the unlimited energy source within me.

MOUNTAIN DEVIL

(Lambertia formosa)

All we need is to imagine our ability to love developing until it embraces the totality of men and of the earth.—Teilhard de Chardin

Lambertia is an Australian genus with nine species, eight of which occur in southern Western Australia. Mountain Devil, being the ninth, is found in sandy soils along the New South Wales coast and in the Blue Mountains.

It is a stiff, upright shrub, 1 to 3 metres in height, with sharp-tipped, glossy green leaves. Like many of the other plants in the Proteaceae family, it can survive long periods of drought. Its flowering time is mainly spring and summer. The bloom is composed of seven tubelike flowers enclosed by long red bracts. Honeyeater birds are often seen around this plant, feasting on the flowers' bountiful stores of nectar.

The plant's generic name honours the botanist A. Lambert, and *formosa* means beautiful—an apt description. Its common name comes from the distinctive double-horned fruits which resemble a devil's head. The fruits are green when young and dry to a dull brown, often remaining on the stem for many years. When the woody fruit finally splits, usually after fire, it releases two-winged seeds.

Mountain Devil is a most important remedy which helps to bring about unconditional love and acceptance as well as forgiveness. It is the key remedy for allowing universal love to come through.

Mountain Devil can be used for all states in which there is a lack of love. Love is the ultimate life force—the strongest energy there is. Our spiritual evolution is geared towards the ability to express greater and greater amounts of love. The fruit of this plant symbolises that evolution, that struggle, for the devil represents a lack of love. Jealousy, envy, suspicion and anger all encompass an absence of love towards humanity.

Many great teachers throughout the ages and from all parts of the world have preached the importance of love because it is the basic foundation for all human relationships. What we do to others is reflected on to all of humanity. To hate another is to hate ourselves, for we are all living within the one Universal Mind. What we think about another we think about ourselves. To let go of all bitterness and resentment and to love one another is what we owe ourselves and humanity.

As this remedy clears away the hatred, anger and other feelings blocking the expression of love, a deep sadness (the other side of anger) may be revealed. This reflects a lack of self-love, another reason why these people have trouble expressing love for others.

The bush essences are self-adjusting, and even when working on very powerful emotions such as hatred and love, they will release only those feelings and show people only those aspects of themselves that they are ready to deal with. Yet the Mountain Devil remedy works at a very deep level and continues to help people even after they have finished the drops, still clearing away the hatred and anger and letting more and more of the love inherent in them come through. Problems with resolving such strong emotions are extremely rare with the bush essences but if a problem did occur the person could simply stop taking the remedy and begin to take Emergency Essence.

Negative condition

•

hatred

•

anger

•

holding of grudges

•

suspicion

Positive outcome

•

unconditional love

•

happiness

•

forgiveness

The personality of the Mountain Devil type is very suspicious and wary. These people don't trust others and guard themselves against being ripped off. Yet charity and goodwill towards others is invariably returned to enhance the giver's own life in so many wonderful ways.

My grandmother used to say that if you can't think of something good to say about a person, don't say anything at all. Yet people with this type of personality are so malicious that often if they can't think of something bad to say about a person, they don't say anything at all.

Associated with their anger is a lot of blaming. They tend to blame the people around them and don't take responsibility for their own feelings or for the circumstances of their own lives. They also tend to see the ugliness of life by projecting on to the world their own frustrations and anger. They can be very spiteful, embittered people who are unpleasant to be around.

However, their own feelings are, in fact, poisoning these people. This can be seen in the iris by the brown toxic colon and the disturbed liver.

As these people are poisoning themselves with hatred and resentment, they may develop illnesses. These often take the form of cancer, strokes or arthritis. The gnarled joints of arthritics sometimes reflect the internal shape of these people. The sharp stabbing pains of the disease are indicative of the barbs they project on to others. These people often hold grudges that stay with them decade after decade, slowly eating away at their bodies and creating illnesses within them. The hatred we harbour for others gives them power over our own health and well-being. Our hatred does not hurt our enemies at all. It only turns our own days and nights into hellish turmoil.

In fact, people who feel a lot of anger and hatred often injure themselves and others in "accidents". Car accidents are a more legitimate way of expressing anger in our culture.

Mountain Devil works very well for people who are involved in a divorce or separation, when a lot of hatred and revenge and manipulation often occurs. "I feel as though my good side has returned" is a not untypical comment from those who have taken the remedy.

The positive aspects of this remedy are the powerful human virtues such as love and forgiveness. This ultimate form of love was shown by Christ with his final words on the Cross: "Forgive them for they know not what they do."

People who are able to give without wanting anything in return and can rejoice in others' good luck embody the positive aspects of Mountain Devil. In the business world there are changes occurring. Instead of the competitive win/lose attitude, a new philosophy is emerging, involving the win/win situation which leads to greater cooperation and an improved quality of life.

It is important for Mountain Devil types to recognise that they are feeling angry and to acknowledge to themselves that they have those feelings. A more effective way to let out anger than smashing something, someone or yourself up is to kneel in front of a stack of pillows and, with a stick raised over your head, hit into them hard, releasing the energy in your spine. You have let your anger out without hurting anyone. If that is too artificial for you, release the anger by doing something physical like digging a hole in the ground or running around the block.

On that score, a child who was exceptionally aggressive and violent would describe in vivid detail past life scenes in villages in Eurasia, where he raped, murdered and pillaged as a member of a horde. His behaviour changed dramatically after he took Mountain Devil. He became much more gentle, kind and tolerant towards his parents and younger brothers.

Mountain Devil has given many people a greater understanding of their anger and creative ways to release it. These people take more responsibility for their actions and blame others less often.

This is a wonderful remedy for jealousy and especially sibling rivalry. When a new brother or sister is about to be born, giving the child some doses of Mountain Devil can help to remove any jealousy. In that situation this remedy is very good for pets, too.

When people set up house together and bring their own animals with them, a tense situation may develop. From a metaphysical point of view, if you visit someone whose dog snarls and barks at you, this is a reflection of the fact that the person doesn't like you. Maybe you could offer that person some Mountain Devil as a peace offering. It is a great eye-opener in a relationship when one person's animal starts barking and snapping at the other person.

The Doctrine of Signatures is quite wonderful in this plant. The red of the flowers indicates the intensity of the emotions. The seven flowers symbolise spirituality, as this number represents the learning of lessons and, of course, the most important lesson to learn is the expression of love. The sharp leaves of this plant are like barbed words that hold others at bay and inflict pain. The double-horned fruits resemble a devil's head. The seeds often need fire to be released. Fire burns off the dross to expose the core. The two-winged seeds symbolise the angels who came through the fires of hell. The fruit remains on the tree for years, like old grudges and resentments which require a cleansing fire to burn them away.

I am now open to universal love.
I am forgiving in all situations now and always.

MULLA MULLA

―――――

(Ptilotus atripicifolius)

*F*ear *is not of the present but only of the past and future which does not exist.*—A Course in Miracles *(eds Frances Vaughan & Roger Walsh)*

The *Ptilotus* genus is commonly known as Mulla Mulla and inhabits the dry inland parts of the Australian continent. There are over 100 species widely distributed throughout the mainland, all but one being exclusively Australian. Like most desert plants, they come into flower when seasonal conditions best suit them. The flowers may persist for months until broken up by wind and rain. Most of the species occur in Western Australia and the Northern Territory, where they can cover hectares of red sandy gravel plains.

Ptilotus atripicifolius is a small annual or perennial with pinkish flowers. The flowers, which are about 2 centimetres long and covered in downy hairs, are clustered in densely packed terminal heads. This plant prefers the rocky, loamy soil of hot, arid regions. The greater the heat, the better it seems to thrive.

The Mulla Mulla remedy was made up in Palm Valley in the Northern Territory, where some of the oldest plants in Australia grow. Not surprisingly, this plant is found in the desert in central Australia, the hottest part of the continent, as its properties are closely related to heat and fire. This remedy is for physical and/or emotional recovery from shattering experiences caused by burns, hot objects and fire.

This remedy is an example of the way in which the bush essences have been provided to meet society's needs at this time. The continuing destruction of the ozone layer is allowing more ultraviolet radiation to penetrate the earth's atmosphere, leading to a higher incidence of sunburn and skin cancers, even in areas where these conditions were previously quite uncommon, such as Argentina, New Zealand and Tasmania.

We have been absorbing much more radiation in recent years, and the Mulla Mulla remedy can help to release this stored radiation from our bodies. After taking the Mulla Mulla essence, many people have experienced old sunburn, sometimes dating back years, appearing on their bodies and lasting for only a day or two. They have been able to determine how long the radiation has been stored in their bodies by the widths of strap marks, which indicate the fashions of various periods. It can combine with Fringed Violet for those having radiation therapy.

The fear of heat and fire is often unconscious and may be manifested in a lack of vitality, as if these people want

Negative condition

•

fear of flames and hot objects

•

trauma associated with fire and heat

Positive outcome

•

rejuvenation

•

feeling comfortable with fire

— 110 —

to fade away. When a person presents with these symptoms, appropriate counselling can often reveal the underlying fear of fire and hot objects.

One patient had been putting on weight for five years; his energy was low and he felt very heavy and sluggish. Muscle testing revealed that he needed Mulla Mulla. It emerged that five years earlier his house had been burnt down in the Ash Wednesday fires of Victoria, in which many people had lost their lives. He had fought the fires and had some terrifying experiences. Ever since then he had suffered from stress in hot weather and had steadily gained weight which had been very difficult to lose, as it was a protective layer. Mulla Mulla was of great benefit to this man.

When we were visiting Rotorua in New Zealand, my daughter, Grace, developed a high fever after being in one of the thermal pools. Again, on muscle testing the remedy indicated was Mulla Mulla. My feeling is that the hot water and fumes triggered a deep-seated memory from one of her past lives.

During past-life regression work, many health practitioners and counsellors have found that the Mulla Mulla remedy is indicated for large numbers of their clients, as these people have experienced being burnt at the stake in former lives. Interestingly enough, many of these people now work in healing-related fields, as herbalists and healers, usually women, were burnt at the stake in an attempt by the Church to gain power.

We only have to look back at the ghastly tortures of times gone by to realise the many ways in which hot instruments have been used throughout history. Thus it is not surprising that traumatic experiences caused by heat and fire, especially in a recent past life, can affect people strongly to the present day.

I now face life with vitality and confidence.
I now release my fears.

OLD MAN BANKSIA

(Banksia serrata)

A thousand-mile journey begins with one step.—Lao-Tse

One of the first Australian plants discovered by Sir Joseph Banks, Old Man Banksias are found in all States of eastern

Australia. Growing as high as 17 metres, as they age they develop stout, gnarled trunks encased in dark, furrowed, pebbly bark. The stiff, glossy, dark green leaves, which grow up to 16 centimetres in length, are evenly serrated.

To observe the full evolution of the tree's flowering cycle is to delight in nature. In late summer the erect, velvety brown buds slowly reveal the immature, silvery green flower spikes. Starting from the base and working up the cone, the flowers gradually open, changing in colour through shades of golden fawn, to orange and red as they wither. Finally, all that is left are the "big-eyed", woody cones thickly coated with soft, short, dove-grey to rusty red fuzz. These cones, which are 8 to 16 centimetres long and contain the winged seeds, are retained for many years and are a striking feature of the tree. The late May Gibbs based her famous characters, the big, bad Banksia Men, on the appearance of the cones. While in bloom, the cones contain vast amounts of nectar which attract many birds, marsupials and insects.

Like other members of the Proteaceae family, Old Man Banksias are very adaptable and can be found hugging barren, inhospitable areas, but more commonly grow in woodlands and open forests.

This is a remedy for solid, plethoric, heavy people who are often low in energy. They are slow-moving and have been sluggish for a long time. They may have been disheartened by setbacks or may have merely felt tired for a long time. The Old Man Banksia essence brings a spark back into these people's lives. It also helps to give them staying power. It continues to act in the body long after a person has finished taking the remedy.

These people often suffer from low thyroid activity, which produces sluggishness, physical heaviness or obesity, and tiredness. They tend to be reliable, dependable people who steadily plod on, often hiding their tiredness and battling on with unceasing effort.

The Old Man Banksia type certainly isn't as fast-thinking as the Black-eyed Susan person. They tend to have earthy natures and operate more from the emotional and physical planes than the mental plane.

They often have a strong affinity with children and may be the main focus of their families, taking on most of the problems and workload. They give a lot of themselves and their time to others. They often suffer from overload and overwork and find it difficult to say no when others want to rely on them.

As well as the sluggish thyroid pattern, these people may suffer from heart attacks or strokes or they may physically collapse or have nervous breakdowns. Old Man Banksia is a good remedy for helping them realise their limits and learn to say no.

These people have a lot of common sense. They don't rush through things, they are practical, methodical and very patient. They are usually very good listeners, as they feel a lot of concern for others, and people in trouble often turn to them. They like helping others and are usually very people-oriented.

These people often have a very strong intuitive sense. They may be psychic and in touch with nature.

One of my patients was involved in many meditation circles. As she was a very good medium, people would always ask her to come to their groups to assist in the members' development. Because she was very aware, she felt compelled to help others. She began to put on weight and feel tired and sluggish, but couldn't refuse these requests. After taking the Old Man Banksia remedy, she started to recapture her vitality and began to say no, and the crippling pain in her knee and hips disappeared. When taking this essence, some people have experienced unusual sensations such as tightness

Negative condition

•

sluggishness

•

plethoric

•

low in energy

•

disheartened

•

weary

•

frustrated

Positive outcome

•

enjoyment of life

•

energy

•

enthusiasm

•

interest in life

in the throat, but these are only temporary and indicate that the energies of the thyroid glands are being rebalanced.

In some Western countries it has been estimated that over 40 per cent of women over the age of forty have underactive thyroids. Old Man Banksia can be of great benefit to women suffering from this pattern. In fact, the Australian Aborigines saw this tree as a symbol of female spirituality. It can be combined with Macrocarpa and Crowea to produce a very strong tonic for vitality.

The positive aspects of Old Man Banksia are enjoyment, enthusiasm, energy and an interest in life, as well as the ability to cope with whatever life brings.

I now cope with all aspects of life.
I now feel joy and enthusiasm for life.

PAW PAW

(Carica papaya)

Here's a two-step formula for handling stress.
Step 1: Don't sweat the small stuff.
Step 2: Remember, it's all small stuff.
                    —Anthony Robbins, Unlimited Power

The Paw Paw is an evergreen tree, growing up to 8 metres tall and crowned by deeply lobed, palmate leaves up to 60 centimetres across. The fruit is a large berry from 10 to 50 centimetres in diameter, arising from the succulent, white, female flowers.

Originally from Mexico, the tree is now native to the tropical regions of America, India, Tahiti, the Malay Archipelago and northern Australia.

The Paw Paw tree has a very straight, narrow trunk which branches into a mass of foliage and fruit at the top of the tree. This is a clue to some of the plant's properties.

Paw Paw is a very good remedy for activating an awareness of the Higher Self, which has the answers to all our questions. When a person is struggling with a major life decision, this is the remedy to recommend. It strengthens the intuitive process and thereby helps us find solutions to our problems. The effect is even more powerful if the person goes into a quiet meditation after taking this essence. In that quiet space Paw Paw increases communication with the Higher Self.

Swamp Banksia

Bauhinia

Billy Goat Plum

Black-eyed Susan

Bluebell

Boronia

Bottlebrush

Bush Fuchsia

Bush Gardenia

Bush Iris

Crowea

Dagger Hakea

Dog Rose

Five Corners

Flannel Flower

Fringed Violet

Grey Spider Flower

Hibbertia

Illawarra Flame Tree

Isopogon

Jacaranda

Kangaroo Paw

Kapok Bush

Little Flannel Flower

Macrocarpa

Mountain Devil

Mulla Mulla

Old Man Banksia

Paw Paw

Peach-flowered Tea-tree

Philotheca

Red Grevillea

Red Helmet Orchid

Red Lily

She Oak

Silver Princess

Slender Rice Flower

Southern Cross

Spinifex

Sturt Desert Pea

Sturt Desert Rose

Sundew

Sunshine Wattle

Tall Yellow Top

Turkey Bush

Waratah

Wedding Bush

Wild Potato Bush

Wisteria

Yellow Cowslip Orchid

When trying to make an important decision, many people feel overwhelmed by the responsibility of choosing the best course of action. The Paw Paw remedy will instantly resolve the feeling of being overwhelmed. It is one of the fastest-acting bush essences.

It is also helpful when a person has been exposed to a lot of new information or ideas that are difficult to grasp. Paw Paw will bring about an assimilation and integration of the information or ideas. It is a wonderful remedy for students and, come exam time, it is used by the bucketload. It is also good for students who have only a few weeks to study for an exam. Because the task is so daunting, they often feel overwhelmed and are reluctant even to begin. Paw Paw resolves that feeling and allows them to take the first step.

As Paw Paw is such an acute remedy, it can be taken in one-off doses whenever a student, or anyone else for that matter, is feeling overwhelmed or is having difficulty in integrating new information. Just before or after a seminar or workshop, a single dose may be taken to integrate the new material.

This remedy was made up during a residential Touch for Health workshop. Everyone was expecting a leisurely, carefree time, as they were told that most of the teaching would be done down on the beach. Many of them were very tired after organising to have two weeks away from their busy practices.

Negative condition

·

feeling overwhelmed

·

unable to resolve problems

·

burdened by decision-making

Positive outcome

·

improved access to Higher Self for problem-solving

·

assimilation and integration of new ideas

·

calmness

·

clarity

Joan and Bruce Dewe decided that they were going to use the workshop as a trial run for a new program they wanted to introduce at the American National Conference. But instead of a leisurely time at the beach, everyone was confronted with having to absorb eight hours of information in only six hours a day. Their heads were spinning at the end of each day. It was very appropriate that a remedy like Paw Paw had its genesis in such an environment.

On a physical level, Paw Paw is used as a protein digestant. As a flower essence its action is very similar, but it works on different levels. Early research has shown that real benefits can be derived from the Paw Paw essence in situations where quality of food intake is diminished, or where there is illness due to the malabsorption of food. It combines very well with Crowea for all digestive or abdominal disorders.

I am now gaining greater access to my Higher Self.
I now easily assimilate and integrate new information.

PEACH-FLOWERED TEA-TREE

(Leptospermum squarrosum)

Life's a pretty precious and wonderful thing.
You can't sit down and let it lap around you . . .
You have to plunge into it, you have to dive through it.
                    —Kyle Crichton in A Bag of Jewels
            (eds Susan Hayward & Malcolm Cohan)

The forty native species of tea-tree are from the same family as gum trees and bottlebrushes. The Peach-flowered Tea-tree is found in the bush areas of the Central Coast and South Coast of New South Wales, as well as in the Central and Southern Tablelands.

The shrub bears abundant large, pink blossoms, about 16 millimetres across, with five petals. The flowers later turn white. The shallow carpel in the centre of each flower contains nectar. The sweetly scented leaves were used by early sailors to make a tealike drink. They are narrow and sharply pointed. The name *Leptospermum* refers to the tiny seeds found in the fruit, which is a hard capsule.

This tea-tree grows up to 3 metres high, preferring areas with a reasonable rainfall, though even during droughts it will flower well. It is often found in sunny, exposed areas

like heathlands and sand-dunes. The plant is rather ungainly in appearance, with the exception of its attractive flowers.

The Peach-flowered Tea-tree is a good remedy for those who lack the will to follow through and for those who have initial enthusiasm that eventually drains away, leaving them without interest. The Jacaranda remedy is not so much for people who lose interest, but for those who lack focus.

The origin of this essence is quite unusual. I had been invited to appear on a television talk show and was asked to bring some of the bush flowers along, to be used as a backdrop. I had seen the Peach-flowered Tea-tree in flower and had been very attracted to it, but I didn't sense its properties. Yet, while gathering flowers for the television program, I kept seeing an image of the interviewer's face superimposed on the flowering tea-tree. I also received a message about one of the properties of the flower. This interviewer had a reputation for extreme mood swings. In fact, I had been warned that if I went on one of his bad days, the interview could be very difficult.

While waiting to go on air, the interviewer complained to me about how difficult shift work was and how his health had been adversely affected by the irregular hours that the industry demanded. He showed me a photocopy of an article about the detrimental effects of shift work.

This remedy is for people who have extreme swings of mood and also for hypochondriacs—those who constantly worry about their health. By focusing on ill health, these people often create the conditions they fear. For example, in the United States at the moment, many people are visiting AIDS clinics with all the symptoms but without the disease. This illustrates the extent of people's concern about contracting AIDS and their preoccupation with themselves and their health, along with the guilt associated with sexuality in our culture. Peach-flowered Tea-tree is excellent for such personal concern, whilst Sturt Desert Rose is the essence to resolve guilt.

Peach-flowered Tea-tree people have vivid imaginations when it comes to their own health and are quite convinced they have all sorts of illnesses. These people are constantly thinking, "What caused this problem, what did I eat, what did I drink?" As one problem is cleared, another will arise. Their worries are all related to their bodies, whereas Crowea people tend to worry about things generally. This essence fosters a balanced and responsible attitude towards one's own health, rather than a preoccupation with it.

Sometimes these people are also afraid of getting old and, if so, Bottlebrush combined with this essence could be helpful. Peach-flowered Tea-tree, Dog Rose and Little Flannel Flower (to activate the child within) also make a good combination for those who fear old age.

For any person who fits the picture of this essence, sunshine is very important. Being in the sun, even for a few minutes three times a day, can help to correct the imbalances within them. Such types will suffer if they have to spend most of their time indoors. They will feel better if they can be near a window, close to light.

As mentioned earlier, this remedy is for those who become very enthusiastic and then, for no apparent reason, lose all interest. It is also good for people who do not follow through with their plans. It allows them to develop stability, consistency and drive. The main reason why they don't complete projects is that boredom sets in. They are quick to pick things up and are good at what they do, but they get bored very quickly.

The change in colour of the flower itself indicates its ability to help with mood changes—swings between joy and depression. When combined with She Oak, it is helpful for premenstrual tension. Peach-flowered Tea-tree has a balancing effect on the pancreas, and one of the functions of this organ is to control blood sugar levels. It also helps to regulate the kidneys, which control the amount of insulin that the pancreas can provide. With kinesiology one can readily demonstrate the quick action of this remedy on the pancreas.

Negative condition

•

mood swings

•

lack of commitment and follow through with projects

•

hypochondriacs

•

easily bored

Positive outcome

•

emotional balance

•

follow through with projects

•

trust in and responsibility for one's own health

— 118 —

The positive aspects of this essence are emotional balance, self-confidence, an ability to achieve goals and taking responsibility for one's own health without being preoccupied with it. People who can benefit from this essence miss many opportunities in life and waste a lot of time and energy. They often become depressed about their own inconsistency and feel frustrated at failing to reap the rewards of their initial hard work.

I now feel a sense of purpose and conviction.
I am now confident about my emotional and physical health.

PHILOTHECA

(Philotheca salsolifolia)

Watch how a man takes praise and there you have the measure of him.—Thomas Burke in 20th Century Quotations *(ed. Frank S. Pepper)*

This shrub is a native of Queensland and New South Wales and is one of five species in a small genus. It grows mainly in heathlands, preferring sandy soil.

Philotheca grows up to a metre high. It has starry, pinky mauve flowers appearing at the ends of the many narrow branches. The five petals, each of which is 7 to 12 millimetres long, open wide around the centre of the flower. The ten straight stamens join at their bases. The leaves are small, narrow and crowded. Philotheca flowers in spring and summer.

I had a rather unusual experience with this essence. At the time I thought I was making up the remedy with *Eriostemon*, a very similar plant. When I got home and did some more research on the plant, I realised that it was Philotheca. This was very appropriate because the main property of Philotheca is the acceptance of acknowledgement, and while I was making up this essence it didn't even tell me that I had mixed it up with another plant. In Philotheca's natural environment the flowers are very easily overlooked.

This remedy allows people to accept acknowledgement for their achievements and for who they are and to let in love from others. Doing this might not sound very difficult. But in Australia, and in many other countries for that matter,

we suffer from what is known as the tall poppy syndrome, whereby people criticise high achievers and try to take them down a peg or two. This insidious syndrome can be seen in schools, where successful students and athletes who express pleasure at their hard-earned achievements are called wankers or boasters. Thus children learn at an early age that mediocrity is preferable to success, that they should never aim too high and that they should be suspicious of those who do.

However, it is very important to set ourselves goals and to accept acknowledgement for our achievements as this helps us remain focused on our goals and projects, which in turn allows us to fulfil our life purpose. In many cases the goals we set ourselves are simply steps leading us towards our main purpose in life.

Philotheca people are often generous, giving and good listeners, but they have trouble with acknowledging themselves. They have difficulty in accepting compliments and tend to be shy.

The Acknowledgement Process

In our workshops we often run the Acknowledgement Process which gives people the opportunity to enjoy the experience of telling others about what they like about themselves and what they feel good about having done. This process is practised with two people who sit opposite one another. Sometimes they hold hands. Eye contact should be maintained during the whole process. One person starts by asking the other, "What do you acknowledge yourself for?", to which the partner answers with an acknowledgement of one aspect of himself or herself. The person just nods or smiles without

Negative condition

•

inability to accept acknowledgement

•

excessive generosity

Positive outcome

•

able to receive love and acknowledgement

•

ability to accept praise

commenting and repeats the question. This continues for a couple of minutes, and if, during this time, the partner can't think of an answer, he or she can simply say "Blank", and the process continues. Then the process changes and the partner who was answering the question remains silent and listens to the other person who acknowledges the partner for the things that he or she said and anything else that can be perceived about the partner. Again, the partner should not comment about what is being said. This continues for a couple of minutes. Then the roles are reversed and the whole process is repeated.

It doesn't matter whether you do this process with a relative or someone you have never met before. Even strangers will use their intuition accurately about you. The beauty of this process is that you can acknowledge whatever you want to about yourself and your partner, and no comment will be made. After this process is finished, most people want to sit and talk about the many issues that have arisen. This is particularly so for couples and for parents and their children, especially when these children are adults. This process is very effective for relationships in which there is a lot of tension, as it allows many feelings and views to be expressed. Recently, when we were running this process in northern Tasmania, a man in his sixties, who had been a farmer since he was fourteen, told his wife for the first time that he no longer wanted to farm. She had felt this was the case, but he would never talk about it. During this process he was able to open up and discuss his feelings. By maintaining eye contact throughout the whole process, the participants keep in touch with their hearts and their deeper emotions.

After this process many people ask for a dose of Philotheca. It becomes evident how well or badly you are able to acknowledge yourself and accept the acknowledgement of others. Even advertising exploits the idea that it is better not to boast, as can be seen by the "quiet achiever" commercials. So it is easy to see how these stereotypes become enshrined in our psyches.

People who are unable to accept praise or acknowledgement or help from others will often become dependent, relying on the help of others perhaps because they become physically incapacitated to some degree. It is so much easier just to let in acknowledgement along the way. Others also feel better when they acknowledge a person who appreciates what they are saying.

Kristin was doing a lot of work with Lloyd Rees, the great old man of Australian art. For his ninetieth birthday a number of dinners were given in his honour. The day after one such dinner he told her she would have to go home as he felt so sick. He explained that listening to all the wonderful things they said about him nearly killed him. I was amazed that such a gifted man who had achieved so much was unable to accept acknowledgement.

So many people receive acknowledgement only on the day of their funeral, and by then it is too late for them to know what others think of them and to let in the love that others feel for them. The power of acknowledgement is very deep, and this remedy helps people open up to that energy.

I now acknowledge myself for who I am.
I am deserving of the love and support offered to me and am now gracefully accepting encouragement and praise.

Red Grevillea

(Grevillea speciosa)

The impossible is possible when people align with you. When you do things with people, not against them, the amazing resources of the Higher Self within are mobilised.—Gita Bellin, Amazing Grace

This is one of the most decorative of the grevilleas, with stunning red flowers that bloom throughout the year, but predominantly in spring. Its specific name *speciosa* means showy. A bushy shrub growing up to about 2 metres high, it is found in the standstone areas north of Sydney.

The flowers form large, circular clusters which occur either at the ends of the branches or amongst the leaves. The tubular red flowers roll back their petals to reveal a single club-tipped style. Passing breezes play with these curved styles, giving the flowers a spiderlike quality.

Many red-coloured flowers are pollinated by birds, whose eyes are particularly sensitive to the longer wavelengths of the visible spectrum, namely, red and orange. The vision of most insects, especially bees, is shifted towards the shorter wavelengths of light. They are able to see ultraviolet but can't see red. Red flowers rarely have any scent as the birds that pollinate them have virtually no sense of smell.

Red Grevillea is a very powerful remedy for people who feel stuck. It gives strength to those who know they need to move on from situations that aren't good for them, but who can't work out how to do so. Sometimes these people depend too much on others and do not use their own resources as well as they could. They are often very sensitive to criticism, which drives them further into themselves. This remedy will

Negative condition

•

feeling stuck

•

oversensitive

•

affected by criticism and unpleasant people

•

too reliant on others

Positive outcome

•

boldness

•

strength to leave unpleasant situations

•

indifference to the judgements of others

help them come out of their shells as it promotes independence and boldness. Homoeopaths have found that this essence can drain golden staph, a bacterium very resistant to most antibiotic treatment, often contracted in hospitals.

If recommending this remedy to others, don't be too attached to your own expectations of what they will do in response to their situations. The steps they take may appear to be backward, but often they need to step around a problem in order to go forward. By going around the problem, they learn the lessons that are important for them and that, in fact, affect the very core of their beings. To an outsider, a more direct path may seem quite obvious, but this may not be in the best interests of the person taking Red Grevillea.

This remedy is extremely effective, even though the results may not be as anticipated. The spiderlike appearance of the flowers hints at the predicament of the Red Grevillea type, who is like an insect caught in a web.

One woman was still married, but the relationship had ended long ago. Her husband lived interstate and paid her a weekly salary. She wanted a divorce but was worried about how she would cope financially once the marriage was legally dissolved. After a week of taking Red Grevillea, she decided to go ahead with the divorce. On the same day her husband rang her and suggested a divorce, offering her a very lucrative settlement—far more than he was legally obliged to give her.

Red Grevillea can be appropriate for many situations. People may be in jobs they don't like, but they are afraid to leave because they have mortgages. Others may be looking after sick relatives, but they feel trapped. These people are in tune with what they need, but they can't work out how to change their situations.

After taking this essence, many people report that bizarre "coincidences" have occurred, which has assisted them in moving on. This is indicative of the way one's reality can change once there is a shift in one's attitudes or awareness.

I am now finding the strength and courage to leave unpleasant situations. I am a powerful, poised individual acting out of strength and courage.

RED HELMET ORCHID

(Corybas dilatatus)

It is a wise father that knows his own child.—Shakespeare, The Merchant of Venice

This orchid is found in all States except Queensland and the Northern Territory. It likes damp, sheltered, mossy or leaf-mould areas under ferns in open forests. Like most of the orchids of south-western Australia, *Corybas dilatatus* is terrestrial, being a dwarf plant with a single heart-shaped leaf growing above the flower.

There is a unique story behind the making up of this essence. In meditation I received a message about an orchid which has the property of being able to help fathers bond with their children. It helps them place less emphasis on their work and become more aware of the importance of nurturing their bonds with their children and spending time with them.

There are many beautiful orchids in south-western Australia, and we often imagined we had found this particular one. But when I tuned into each plant, it didn't have the required property. One day we were at the foot of Toolbrunup Peak, in the Stirling Range. A dirt road ends at the car park, from where a walking track leads one to the top in four or five hours. We had parked and were photographing some plants about half-way up the road, when a car pulled up

Negative condition

•

rebellious

•

hot-headed

•

selfish

Positive outcome

•

male bonding

•

sensitivity

•

respect

•

consideration

— 124 —

opposite us. An excited man jumped out, rushed over and said, "You like flowers, don't you!", to which I merely nodded my head. He went on to tell me about a beautiful orchid, the Red Helmet Orchid, which he had seen in flower. While he was talking it began to dawn on me that the Universe was conveying to me the location of this flower.

I searched for an hour along the walking track. I had been so excited that I had failed to listen carefully to the man's directions. He had said that it was very easy to miss, as it was so low on the ground. As I passed the occasional person returning, I asked them if they had seen the orchid. A few said that the same fellow had also told them about it. One said that it was growing in a mossy area with a lot of leaf mould. Then I came across a spot that fitted this description quite well. I looked around, expecting to see a vibrant red flower, but I couldn't find it. Then, just as I was about to leave, a couple who had also spoken to the man came along. They showed me the orchid. I would never have found it without their help. It seemed such an insignificant flower for one with such potential.

On tuning into the orchid, I discovered that it could not only help fathers with the bonding process, but it could also assist children and adults who have problems with authority figures. Such people have very rebellious natures which can often be traced back to a poor relationship with their father. Even when the father has died, this bush essence helps them

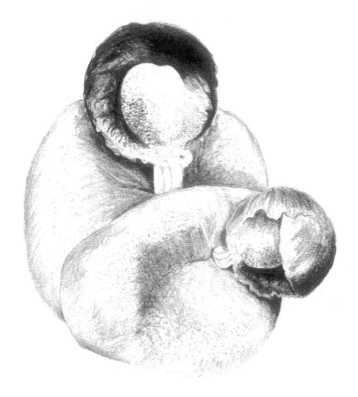

remove the emotional blocks caused by that relationship. Thus it also helps them resolve their negative feelings towards authority and men in general. Many individuals would agree that their relationship with their father was unsatisfactory. No wonder Ashley Brilliant, the American satirist, commented tongue-in-cheek: "Parents and their children are not natural friends but in certain cases lifelong attachments have been formed."

Toolbrunup is a very wind-swept, rugged place, and there was some rain that day. However, the flowers were in sunshine, about half-way up the mountain. I continued my climb to the summit, and by the time I came back the flowers had been steeped for over three hours. But it seemed as though the essence still wasn't quite ready. When I tuned in again, the message was that it needed not only sunshine but also moonlight. Trusting the process, I left the essence there and returned to the car. Then I realised that it was the night of the full moon and that the essence also needed the feminine energy of the moon for its bonding property. It needed both the female principle and the masculine principle of the sun.

The first person who took the Red Helmet Orchid essence had been having great difficulty with a male neighbour, who was very angry and abusive. This had created great fear and anxiety in her. When she discussed this man, it appeared to me that he embodied many aspects of her father. After taking the essence, she had a conversation with her father and found she was able to relate to him in a different way. Before that time she had always found it difficult to speak to her father. Now she was able to express her feelings, and his response to her was quite different. Corresponding with these changes, she found she was able to deal with her neighbour, expressing her frustrations and not being so paralysed by fear.

The following letter from another woman also provides a good example of the problems with authority figures that stem from a poor paternal relationship:

Red Helmet was given to my brother, who could not get along with schoolteachers or principals and was forever in trouble with the police—he drove a motorcycle without a permit and was under age. Now he has a good job, communicates well with Dad and his boss, and no longer gets into trouble with the police. Thank you very much for discovering the Red Helmet.

This essence helps people find their own authority within.

In Western Australia you need a permit to pick wildflowers, and we had obtained one before our trip. The back of the permit contained a list of plants which you are not allowed to pick, even with a permit. As I returned to the car, I noticed that the *Corybas* species was on the list. I wasn't sure if this included the Red Helmet Orchid. It seemed to me that if it did, then picking the flower would be an illegal act, and consequently the remedy made from it would not work. However, I later discovered from a government department that it can be picked. This illustrated the plant's connection with the principle of authority. Moreover, as Kristin was seven months' pregnant at that time, making up this was particularly timely for me.

Many men overlook the importance of bonding with their children, because they are so busy with their work. They tend to be unaware that their children need them as well as their mothers. Metaphysically, a child between the ages of birth and three bonds predominantly with its mother, while from four to seven the father is more important. The Red Helmet Orchid Bush Essence has become available at a time when, for the first time in recorded history, fathers are often present at the birth of their children. This brings up the question of whether or not children are able to bond with both parents from the time of birth. Perhaps research will be done into this subject.

This essence also has the ability to help people become aware of and feel concern for the whole planet. The earth must be treated like a newborn baby because of its fragile environment and the negativity surrounding it. The negative actions of those living on the earth attack not only the earth's surface but also its very soul. If the earth is not treated with great care and gentleness, it will, in turn, attack all life on it and the destruction will be devastating. Red Helmet Orchid can develop our concern for the planet, as well as change the consciousness of men.

I now express greater care and gentleness in my relationships.
I am now able to communicate easily with those in authority.

RED LILY

(Nelumbo nucifera)

If you have built castles in the air,
 your work need not be lost;
 that is where they should be.
Now put the foundations under them!
                    —Henry David Thoreau

Red Lily is a perennial aquatic herb arising from a submerged, creeping rootstock. It is found in the temporary floodplain swamps at the Top End of Australia. The floodplains do not dry out completely after the heavy rains of the wet season, and the Red Lily is usually found in the deep basins and billabongs of the swamps.

This plant is not technically a lily, but is, in fact, a lotus, as the very large, round leaves do not float but rather stand erect and cupped above the water. The stunning, large, prominent, fragrant flowers with deep-pink petals and yellow centres are 15 to 25 centimetres across and can be seen from March to November sitting on prickly stalks a half to one and a half metres tall.

The root tubers and seeds are edible and were used by the Aboriginals. The seeds are extremely long-lived, with some germinating after 200 years.

This plant is also known as the Sacred Lotus. In Buddhist tradition the lotus is a symbol of spirituality. From the murky water arises the pure flower, thereby symbolising the raising of spiritual consciousness. Its distribution from Sri Lanka to northern Australia includes many Buddhist countries.

The Red Lily remedy helps to balance the spiritual and earthly planes. It keeps people grounded and practical, while at the same time allowing them to reach out and touch the spiritual realms.

This essence has also been made up in Bali, where it has slightly different properties from the Australian essence. The latter embodies the properties of Australian Aboriginal spirituality which is very strong and grounded.

Negative condition

•

vagueness

•

disconnectedness

•

split

•

indecisive

•

lack of focus

•

daydreaming

Positive outcome

•

grounded

•

focused

•

living in the present

The negative aspect of this remedy is represented by people who are not grounded on the earthly plane. Some of them are, in fact, resisting being here. They are up in the clouds, daydreaming. They lack an interest in worldly events and the present, and tend to fantasise and live in the future. There is often a vacant, split, distant look about these people. They may be unhappy with their present circumstances, so they escape into a world of fantasy, daydreaming of other situations.

Red Lily people often lack concentration because their minds are elsewhere. They are very absent-minded and impractical, and live in thought rather than action. Their memories are poor because they don't pay enough attention to the events occurring around them. Having conversations with these people is frustrating because they don't really listen to what is being said to them. They will often interrupt and go off at tangents. In this regard, the Red Lily remedy is very beneficial for autistic people.

On the physical level, these people tend to be very clumsy because they don't pay much attention to what they are doing. They may have lots of accidents. They also sleep a great deal, which is another form of escape, and they may be drawn to drugs, especially hallucinogens or marijuana. They are certainly not good people to meet while they're driving.

I used to know a man whose mother would ring him on Sunday afternoons at about 3 pm and apologise in case she had woken either her son or anyone else in the house. Quite often he was still asleep—he would rarely get up before noon. He was a heavy dope smoker and often had a very sleepy, dazed look. He was also a brilliant mathematician.

Red Lily can also be used by people who are suffering from the detrimental effects of taking drugs, especially hallucinogens such as LSD or "magic" mushrooms. Many young people are, as the expresssion goes, "veged out"—they have taken one trip too many and are permanently spaced out. This remedy can also be given to anyone in a coma or to catatonic schizophrenics, who don't interact with anyone at all.

The Botticelli painting *Birth of Venus* is reminiscent of these people, who are often very ethereal and otherworldly.

Red Lily has very similar properties to those of Sundew. However, the Red Lily remedy is for people over the age of twenty-eight, while Sundew is for people up to that age. Red Lily produces a sense of being grounded and focused in the present. It is an excellent remedy for people who have trouble with concentrating and starting on the job at hand and coping with practical tasks.

I am focused in the here and now.
I am now grounded in the physical plane.

SHE OAK

(Casuarina glauca)

*W*hat we vividly imagine
ardently desire
enthusiastically act upon
must inevitably
come to pass.
> —*C.P. Sisson in* A Bag of Jewels
> *(eds Susan Hayward & Malcolm Cohan)*

Casuarina trees are widely distributed throughout Australia. They have a great affinity for water and are used throughout the world for stabilising sandy coastlines.

The casuarina was one of the first trees to evolve on earth. It was named after the *Casuarius* genus of birds, namely, the large, flightless cassowary, whose feathers resemble the tree's long, drooping branches. Its common name, She Oak, came into usage because the early white settlers, who used its timber for furniture, shingles and house construction, regarded it as the poor man's oak.

Casuarina glauca grows in wetlands along slow-flowing streams or beside the backwaters of tidal rivers. It is a tall, straight tree with sparse, weeping branches and pinelike foliage. The female flowers occur in globular heads, with the styles hanging out to catch the wind in a very similar way to that in which the Fallopian tubes wait to catch the eggs ejected from the ovaries. The female flowers grow into nutlike fruits which are the same size as human ovaries. The male flowers are densely packed in long spikes at the ends of the branches, giving the tree a reddish tinge.

For a woman, the inability to conceive can be one of the most shattering experiences of her life. The major function of the She Oak remedy is related to the emotional factors inhibiting fertility. The essence was made up using the female flower of the She Oak. It will benefit women who, for no apparent physical reason, are unable to fall pregnant.

Over 20 per cent of all cases of female infertility are caused by unknown factors. She Oak will clear any conscious or unconscious emotional blocks that may be stopping a woman from conceiving. Feelings of inadequacy or a lack of self-confidence may prevent conception. A woman may doubt her ability to create a baby. She may be worried about how she will cope with pregnancy and child-rearing or about the financial strain of bringing up a child.

Negative condition

•

hormonal imbalance in females

•

unable to conceive with no physical reason

Positive outcome

•

hormonal balance

•

conception

•

fertility

From a naturopathic point of view, many cases of female infertility are due to a dehydration of the uterus, and though elaborate techniques can be used to correct this problem, taking the She Oak essence has exactly the same effect. Simply adding two drops of this essence to a glass of water will allow an individual to properly absorb and use water. Many people are dehydrated, even though they do drink water. It can also be taken to help clear mucus blockages of the Fallopian tubes and create normal healthy cervical mucus. In addition, it will regulate the production of reproductive hormones in women, especially where the ovaries are functioning spasmodically, resulting in an irregular menstrual cycle.

For infertility, the She Oak remedy is used somewhat differently from the other bush essences. It should be taken for a month, and then, after a two-week break, taken for another month. The results with She Oak have been exceedingly good. Of those women for whom I have prescribed the remedy, only one has taken more than six months to fall pregnant, while only two are still not pregnant. However, if a woman does not fall pregnant within six months of taking this remedy, a combination of Flannel Flower and She Oak is recommended. The two essences taken together have the ability to clear karmic blocks that may be hindering conception.

A number of gynaecologists now use the She Oak essence to treat hormonal imbalances, including premenstrual tension. Many alternate therapists also prescribe this remedy for fluid resention associated with the menstrual cycle. In kinesiology, this essence is used to balance the ovaries, and in many cases it will also balance the testes in men.

I now release all emotional blocks preventing conception.
I am now becoming more and more confident about my ability to create.

SILVER PRINCESS

(Eucalyptus caesia)

The great and glorious Masterpiece of life is to know your purpose.—Michel de Montaigne in A Bag of Jewels *(eds Susan Hayward & Malcolm Cohan)*

Eucalypts are often referred to as gum trees, a term coined by Joseph Banks in 1770, when he saw the brown, gumlike sap produced by these trees. Today the term is usually reserved for smooth-barked eucalypts.

The eucalypts dominate the Australian landscape with over 600 species. They are found in most habitats, ranging from the snowline, the Central Desert, rainforests, sclerophyll forests, the mallee and the brigalow.

The name of the *Eucalyptus* genus was derived from the Greek words *eu*, which means well, and *kalyptos*, meaning covered. "Well-covered" refers to the small, hard cap that fits tightly over the eucalypt's flowering bud. The cap has evolved as a fusing of the sepals and petals, both of which are absent from the flower. As the flower blooms, the expanding stamens, which provide the flower's beauty, push off the dried cap. *Caesia* means bluish grey, the colour of the stem and bark of this glorious tree.

Many eucalypts are now exported, not only to be grown for their beauty and for their timber, but also for their ability to stabilise soil and drain swampy land. All gum leaves are leathery, and by hanging down with their edges exposed to the sun's rays rather than their surfaces, they avoid being dried out.

The mallee form of growth is the eucalypt's adaptation to a harsh environment with extremes of temperature and rainfall, bushfires and poor soil. There are almost a hundred species of eucalypts that grow in the mallee form. *Eucalyptus caesia* is one of the most beautiful of all these mallees. This eucalypt has adapted itself to the restricted habitat of granite outcrops in the central wheat-belt areas of Western Australia. It was in danger of extinction some years ago, when it was restricted to one last naturally occurring stand, at Boyagin Rock. Those granite outcrops are believed to be up to 2000 million years old—or half the age of the earth.

However, Silver Princess is now one of the most widely cultivated trees in Australia. It has a slender trunk with pendulous branches and reddish brown bark overlaid with a silvery, waxen bloom. The buds, seed capsules and young branches are also silvery, which is more evident in full sunlight. The bark splits longitudinally into a number of narrow curls. The leaves are silver when young, turning a blue-grey colour. The red flowers grow in clusters of three and dangle on long stems on the swaying branches, unhidden by foliage. Each flower is up to 3 centimetres across, with a fringe of firm stamens around the large, light yellow style.

When I arrived to visit friends in Berri in South Australia, they immediately ushered me to a spectacular flowering red gum growing nearby. I had never seen a more beautiful tree, yet I was unable to find out its name. A number of months

Negative condition

- aimless

- despondent

- feeling flat

- lack of direction

Positive outcome

- motivation

- direction

- life purpose

later, when preparing for a trip to the south-west of Western Australia, I received some channelled information about a red-flowering gum that would be very important for giving people a sense of their life's direction. On arriving in Perth, I saw numerous examples of *Eucalyptus caesia*—my red gum from Berri. Unfortunately, all these trees were growing either near main roads or in environments that weren't suitable for flower-essence-making, or they had finished flowering.

In the Western Australian botanic gardens, Kings Park, one of the botanists told me that the flowering season of Silver Princess had finished and that the only naturally occurring stand of this beautiful eucalypt remaining in the world grew on a granite outcrop known as Boyagin Rock, near her home town of Brookton, three hours south-east of Perth. She described how to get there but said that there was no point in going because the trees wouldn't be in flower. There was, however, a stand in the botanic gardens with just a few flowers left, and I was very tempted to make up the essence from these flowers, even though they were a little too close to a car park. Perth experienced a very violent storm, which blew off the remaining flowers, so I concluded that they obviously weren't meant to be used. Our gut feeling was to go to Boyagin Rock.

At Boyagin Rock we arrived to find the stand of *Eucalyptus caesia* in full, glorious bloom, the red flowers months out of season. I felt that the trees had been kept in flower for me in another example of the divine intervention that has helped me make up these remedies, which have such an important role to play.

It was very satisfying to finally make up this essence. It is for people who are uncertain about their life plan or purpose. Many people never actually know their complete life plan but let it unfold during their lives. This remedy can be taken when they have come to a crossroads or turning point and are not sure which step to take next. At these times, they have no clear goal or direction.

It is such an extremely satisfying feeling to be doing what we know we really need to do with our lives, or at a particular phase of our lives. When we are just drifting along we always feel frustrated by the sense that something is missing.

The education system in most Western countries requires high school students to make a choice about their careers before they know what they really want to do. The earlier these young people begin to meditate, the easier it is for them to find their direction in life. During meditation or in quiet, reflective states we are open to guidance from our Inner Self, which helps us find our life path.

Some people don't become clear about what they want to do until later life, when it can be more difficult to make the necessary changes. However, if they are really clear about what they want to do and are determined to do it, things invariably fall into place and opportunities arise, although they need a certain amount of faith to take the first step. If, towards the end of life, people have not yet found their life purpose, they may feel very despondent, believing that they have wasted a whole lifetime.

Another important role for this essence is to help people find a new direction once they have achieved a major goal. When people focus on an important goal, they often ignore the day-to-day aspects of life, so that when they achieve their aim they are left feeling very flat, with a sense of "Well, so what! Is that all there is?" The Silver Princess remedy allows them to enjoy the journey towards their goal, and once they have reached it, gives them the motivation to pursue new goals in life.

It is very important that we have our goals in life and that we work towards realising them. In many cases the real benefits of these goals are quite different from what we perceive to be their main purposes. We are like the bee that leaves its hive each morning with the purpose of collecting pollen. Yet the main benefit of the bee's activity is to pollinate the flowers and thus complete the natural cycle of plant regeneration.

Similarly, the benefits that flow from our goals are like the ripples that flow across a pond after a pebble has been thrown in. By looking at the consequences of our goals, dreams and desires, we often perceive what our true purpose has been.

Many people notice the effects of taking the Silver Princess Bush Essence within a short time, but be prepared to use this remedy for more than the customary two weeks. Certainly the quality of life for the individual and for society as a whole will improve dramatically when people are able to know and then follow their own life path.

I am now aware of and follow my true life path.
My actions now reflect my life plan and purpose.

SLENDER RICE FLOWER

(Pimelea linifolia)

Skin can be different, but blood same.
Blood and bone . . . all same.
Man can't split himself.
                              —*Bill Neidjie,* Australia's Kakadu Man

All States are represented by the eighty-odd species of the *Pimelea* genus. All are small shrubs—*Pimelea linifolia* grows to about 1 metre high. It blossoms almost all year round. The round, terminal flower head is about 25 centimetres in diameter and consists of a cluster of white, tube-shaped flowers with orange-dotted centres, each with four petals. It has small, narrow, simple leaves which are usually found arranged in a spiral around the stem.

This species is very common in sandy heathlands and low, open forests. Its name is derived from *pimelea*, meaning fatty—referring to the seeds—and *linifolia*, meaning narrow leaves.

The *Pimelea* species were first described by Banks, Solander and Cook. During their brief stay at Botany Bay in 1770, Banks and Solander collected 3000 plant specimens representing over 200 species. As one of the major properties of this plant is cooperation, it is ironic that Banks was involved in having the whole continent of Australia declared *terra nullius*—within the British legal system unoccupied or unused land was able to be claimed by the British Crown. Though the First Fleet settlers found that the Sydney area was occupied by the Aborigines, the British policy was never changed.

The Slender Rice Flower essence helps to bring about group harmony and cooperation and to overcome racism and narrow-mindedness, enabling an individual to see both sides of a question or situation.

During this century the world has experienced many wars, in which over 120 million people have been killed. Many hundreds of millions more have been wounded, have lost members of their families and/or have become refugees. These facts are a grim reminder of our lack of tolerance and love for one another. Past life regression reveals many cases of previous lives where the individuals were involved in wars and believed that their enemies were inferior. Even today the attitude that other nationalities and races are inferior is very common. Many people who lived through the Second World War still hate the Japanese. As with Mountain Devil and other remedies for resentment and hatred, these feelings and attitudes adversely affect the people who harbour them and can lead to serious illnesses.

This plant's structure is very simple. The individual flowers cluster together to form a spherical flower head which symbolises the unity of all human life. This essence helps to release that universal understanding. Christ taught that we should love our neighbour because we are all one. What we do to someone else we are actually doing to ourselves. "You cannot know your own perfection until you have honoured all those who were created like you." (*A Course of Miracles*, edited by Frances Vaughan and Roger Walsh).

The Slender Rice Flower helps people perceive the connection between all human beings and to understand that, by hurting someone else, they are really hurting themselves. As J. Allen Boone expressed it in *Kinship with All Life*: "We are all members of a vast cosmic orchestra in which each living instrument is essential to the complementary and harmonious playing of the whole."

Through the perspective given by this remedy, people can stop making detrimental comparisons and judgements about other religious or national groups. It helps them recognise that we are all at different stages of evolution and that no one is in a position to judge another person or group. This essence helps people to see the divinity and beauty in everything and everyone.

Slender Rice Flower can also help to unleash the humility that is part of this deeper understanding, whereas pride and jealousy represent a lack of understanding. Humility leads to greater harmony and cooperation between people. The concept of synergy means that the whole is greater than the sum of the parts. Thus the cooperative action of a group is much more powerful than the actions of all those people working independently. When people share their resources and knowledge, everyone can benefit. For this to happen group members must be tolerant, flexible and willing to listen actively in order to cooperate for the common good. Slender Rice Flower can help to bring about synergy. This essence is becoming widely used by participants at meetings and

Negative condition

• pride

• jealousy

• racism

• narrow-mindedness

• comparison with others

Positive outcome

• humility

• group harmony

• cooperation

• perception of beauty in others

conferences and also in potentially stressful workplaces such as restaurants, resulting in much higher productivity.

A good example of the success of synergy was seen in the work of the American Edward Denning, who approached managers of the car-manufacturing industry in Detroit after the Second World War, claiming that he could raise their productivity, but they weren't interested. However, he was approached by Japanese car manufacturers who, at the time, had a reputation for substandard production quality and technology.

He realised that for productivity and quality to improve, workers and management needed to cooperate with each other to develop a unity of purpose and a harmonious working environment for all employees, from the managers to the floor sweepers. Most people want to do a good job and to do the best they can. Those who fail usually do so because of poor management systems; rarely is it the fault of the individual. Denning implemented new systems such as quality control, so that employees took pride in their work. Today Japan has a reputation for high quality products with a very low rate of rejection, while in America the opposite situation exists.

A study showed that in 1987 eighty-five films depicting the Russians as baddies were made in the United States, while in Russia fifteen films showed Americans as the baddies. The media is partly responsible for the stereotyping that causes alienation and hostility between nations. This manipulation of the emotions has created the frenzy of hatred that leads to war.

In Slender Rice Flower we have an essence that can help to shift humankind away from nationalistic and religious prejudice and towards global harmony, cooperation and peace.

I now express my love and acceptance of all people.
I now perceive the unique beauty in myself and all human beings.

SOUTHERN CROSS

(Xanthosia rotundifolia)

You are given the gifts of the gods; you create your reality according to your beliefs. Yours is the creative energy that makes your world. There are no limitations to the self except those you believe in.—Jane Roberts, The Nature of Personal Reality

Southern Cross is so named because of the unusual and striking four-rayed compound umbels consisting of clusters of tiny individual flowers held on slender stalks which radiate from the apex of a common stem. These four stems of equal length are arranged on one plane like a cross, with broad, petaloid bracts underneath each cluster of flowers.

There are about twenty species in this Australian genus, most of which are endemic to Western Australia. Southern Cross is a perennial herb found from Mount Barker south to Albany, and we made up this remedy in the Stirling Range.

The major sphere of action for this remedy is resentment and a victimlike mentality. These people feel that life has been very hard on them, that they have been unjustly singled out by fate and that their efforts have been unrewarded. "It's just not fair!" you may hear them say.

With the Southern Cross type there is a total denial of the fact that they are responsible for creating their own reality. They believe that everything is being done to them and that they have no control over their own lives. As George Bernard Shaw said:

> People are always blaming circumstances for what they are. I do not believe in circumstances. The people who get on in this world are the people who get up and look for the circumstances that they want and if they cannot find them, make them.

Each of us creates either a positive or a negative reality. These people create a vast amount of negativity around themselves.

Negative condition

- victim mentality

- complaining

- bitter

- martyrs

- poverty consciousness

Positive outcome

- personal power

- taking responsibility

- positivity

They feel unable to bring about positive circumstances in their lives. They often resent those who are "lucky" and successful. They may blame many people around them but never themselves for their own situations. They may be irritable, sulky or sullen with a whinging manner. They feel that the world owes them a living and had better look after them as they have been victims of circumstances beyond their control. Though these people don't always blame others, they believe that life is against them and always expect the worst.

A woman I know who uses the bush essences regularly became concerned when she was transferred to the Southern Cross High School in Victoria and realised that a victim mentality existed at the school, among not only the students but also the teachers. She treated her experience there as an opportunity to offer some insight into their problems and to increase their positive attitudes and self-esteem.

Numerologically, the arrow of frustration is very common for people with this outlook. If you have a child whose birth chart does not include the numbers 4, 5 and 6, it is very important to support them and help them see the benefits and positive aspects of all their experiences. The following advice will also help: "Feel yourself grow with every experience and look for the reason for it." This will ease their frustration and help them realise that they have a direction in life. It will also encourage them to use their personal power and to create an optimistic view of life.

A positive expression of this remedy is that these people can be very considerate of others. They have been through difficult times and can appreciate what others are going through without judging them. They are able to guide those who feel they are victims of life and show them their problems in another perspective. At times life does seem unfair—the death of a loved one, the end of a relationship, the inability to conceive, money problems—and these people are wonderful helpers at those times.

Also associated with this essence is poverty consciousness. In a wonderful line from the Hollywood movie *Mame*, she cries, "Life is a banquet and all you suckers are out there starving." The Southern Cross remedy helps to open people up to the knowledge that this can be an abundant life. In many countries poverty consciousness is so strong that expensive cars such as Jaguars and Rolls Royces attract immense resentment and hostility. However, if the beauty and craftsmanship inherent in these cars were appreciated, these people would be much more likely to experience abundance in their own lives.

The sense of being crushed and tossed around by life certainly speeds up the aging process, and this combined with bitterness can also affect the internal organs, especially the liver and gall bladder. A feeling of resignation may simply lead to a lack of energy and vitality. Alternating Southern Cross with Sunshine Wattle can be very beneficial for these people.

If you consider the name Southern Cross, it is important to remember that the Cross has "already been borne for you". A helpful technique for the Southern Cross type is to have someone ask you a list of questions to which you respond with the first thing that comes to mind. The person asks you to state the benefits you got from a list you have made for all the seemingly horrendous situations that have occurred in your life or all the things that you feel you have had the least control over in your life. Then you answer: "The benefit I got from . . . was . . . ". You may give some very interesting responses. You can also use this technique on your own with a pad and a pencil. Just write: "The benefit I got from . . . " and add the situation or other undesirable element in your life and the benefit you got from it.

Every difficult situation is an attempt by your Higher Self to show you how you are off-course and to help you make corrections that will bring you back on-course. You always have guidance and spiritual help and need only ask for it.

There is the story of a man who, after his death, saw his life replayed as a series of footsteps in the snow. For most of our lives there are two sets of footprints, the second belonging to Jesus. When life was very difficult for this man, only one set of prints was visible. Finally, he turned around to Jesus and said, "Where were you when I really needed you?" Jesus answered, "I was carrying you."

We are never alone during our struggles as His love is always there to help. We are always capable of dealing with situations, even those that we find intolerably difficult. Our reactions to those situations affect us more than the situations themselves, and every situation is a learning experience.

I am realising more and more that I create my own reality.
I am growing with each life experience.

SPINIFEX

(Triodia species)

All physical disease has its origin in the emotions.—Ian White

There are over thirty-five species of *Triodia,* all endemic to Australia. Spinifex is a native grass which prefers the arid

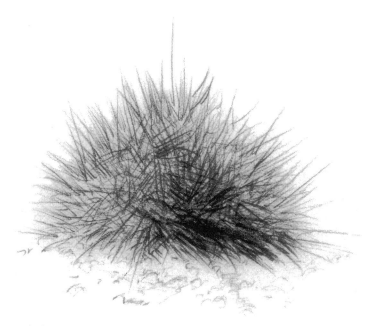

and semi-arid areas of inland Australia. It has adapted to these conditions with its compact, tussocky habit and needlelike leaves which project in all directions. Its growth is centrifugal, and as the central part of the tussock dies it leaves a ring of new grey-green growth.

Spinifex flowers after the summer and autumn rains, creating the illusion in the outback of fields of wheat or a sea of grass as the wind sends ripples through the golden flower heads.

The Aboriginals collected the seed and made it into flour, while the resinous gum of the leaves was used to fasten heads to spears and as a general adhesive. A simple eyewash can also be made by squeezing the juice from the new shoots.

Synergistically, Spinifex has a very powerful, almost masculine energy due to the closeness of the grass to the earth. When I first received the channelled information that I would work with grasses, I was quite intrigued, especially when it was mentioned that the Doctrine of Signatures is more evident in the grasses, being more amenable to the physical senses. By tuning into a blade of the grass, I understood that this essence is relevant to the physical body. The seed heads of these grasses reveal their unique healing qualities far more readily than flowering plants.

Another feature of Spinifex is its green colour. Many people at some point in their lives will need a green essence as well as flower essences. Essences that are made from grasses can be taken internally or can be applied externally to the skin.

Green essences have a cleansing action, clearing parasites, fungal infections, micro-organisms and metabolic waste products from the body. The specific green essence I work with clears monilia and candidiasis, is excellent for toxic colons

Negative condition

·

physical ailments

·

herpes

·

chlamydia

·

fine cuts

·

sense of being a victim to illness

Positive outcome

·

empowerment through emotional under-standing

·

physical healing

and can be used topically to heal all manner of skin lesions, from acne to eczema and psoriasis.

Spinifex heals skin conditions that cause cutting or stinging types of pain, such as herpes. It is very effective for fine, clean cuts where there are no pieces of flesh missing. For these conditions the remedy can be used both internally and externally. For topical application, put seven drops of the essence in clean water, and place a moistened pad over the wounds or fill a misting bottle and spray the lesions. Repeat this morning and night. You can use the essence topically and internally at the same time.

Many people have reported that when the herpes blisters are beginning to develop, spraying the area will abort the attack. This essence will greatly reduce the severity of an attack of herpes and will clear the lesions within a couple of days for most people, whereas without treatment they would remain for at least a week or ten days.

The Spinifex essence can also be used for treating chlamydia, an organism that can cause throat infections as well as infections of the urinary tract and reproductive system. In males the symptom is a pussy discharge from the penis, whereas there is no obvious symptom in females, though chlamydia wreaks havoc in the Fallopian tubes, scarring them badly. In fact, every year in Australia 6000 women become sterile as a consequence of chlamydia, which is often only detected when their infertility is medically investigated.

By taking the remedy internally, Spinifex helps to bring to the surface the emotional causes of blistering skin diseases. The emotional problems can then be resolved with the appropriate flower essences and/or counselling. Many cases of herpes, for example, are caused by sexual guilt, which can be treated with Sturt Desert Rose, or by sexual abuse or trauma, which can be alleviated by a combination of Fringed Violet and Wisteria.

With herpes, for example, sufferers often feel that they are victims of the illness. It recurs repeatedly and for no obvious reason, weakening a person's immune system and confidence and creating physical discomfort or pain. In fact, the attacks are usually triggered by negative beliefs which were formed early in life and which are stored deep in the person's subconscious mind. Many people feel greatly empowered when they understand that whatever happens on the physical plane invariably stems from the emotional plane and that they are able to change the beliefs and attitudes causing their physical conditions and thereby bring about their own healing.

I am now seeing the nature and effect of my deeper emotions.
Increasingly, I am understanding the emotional causes of my physical problems.

STURT DESERT PEA

(Clianthus formosus)

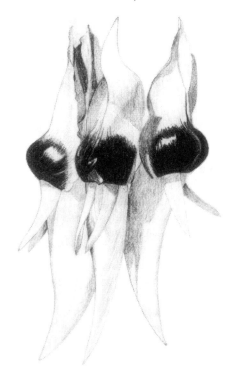

When one door closes another opens. Expect that new door to reveal even greater wonders and glories and surprises. Feel yourself grow with every experience. And look for the reason for it. —*Eileen Caddy,* Opening Doors Within

The Sturt Desert Pea is the floral emblem of South Australia, and as this flower grows in the desert and salt lake areas of the arid interior, it is an appropriate emblem for the driest State in Australia. It is also found on the dry western plains of Queensland and New South Wales and in the arid inland areas of Western Australia, from Kalgoorlie to the Kimberleys.

The Sturt Desert Pea is one of the most spectacular of native plants, possessing tremendous power and presence. Its brilliant, glossy red flowers with shiny dark centres are 7 to 10 centimetres long and form dense heads of four or five flowers.

Its seeds, like many legume seeds from arid regions, have been known to flower after forty years. The longevity of the seeds reflects the main property of this essence, which is for deep hurts that have been stored for many years. The seeds are very hard to germinate. Germination often does not take place until the seeds have been subjected to a great force like fire or boiling water, reflecting the cathartic release of deep hurts.

The plant's generic name, *Clianthus*, comes from the Greek words *kleos*, meaning glory, and *anthos*, flower. *Formosus* means beautiful or well formed. The Sturt Desert Pea was named after the explorer Captain Charles Sturt who, during an expedition from 1844 to 1845, wrote admiringly of the sea of red rising from the bleak, inhospitable wasteland of the arid interior. Yet the first explorer to find the Desert Pea was William Dampier, who in 1699 gathered specimens on the north-west coast of Australia.

Sturt Desert Pea and Waratah are, to my mind, the most powerful of all the bush essences. The main property of the Sturt Desert Pea remedy is that it resolves very deep pain and sorrow. This remedy works extremely quickly in almost all cases, even when the pain has been harboured for many years, even as far back as a previous life. Sturt Desert Pea has brought about some amazing changes in the lives of those who have taken it, especially those who keep their sadness bottled up and rarely cry. After taking this remedy, those people have been able to express their feelings and have felt comfortable about exploring the origin of their pain.

Grief normally affects the lungs, as the following example illustrates. A woman who came to see me had been experiencing deep depression resulting from her daughter's death by drowning. I noticed that her lungs were very tense and prescribed the Sturt Desert Pea remedy. While taking this essence, she cried much more often, and after each crying spell her breathing became easier. She also felt that she was beginning to come to terms with her daughter's death.

The Aboriginal legends based on the Sturt Desert Pea all concern grief, sadness and loss. The following story is an abbreviated version of an Aboriginal legend about how the Sturt Desert Pea came into being:

Long ago the young, strong, proud Wimbaco Bolo eloped with the beautiful Purleemil, the promised bride of mean, old, cowardly Tirtha. They took refuge with a hunting party from another tribe, which was camped near a large lake in the back country.

Wimbaco's father had been born into this tribe, and had been its leader. Realising that the strength of Wimbaco and the beauty of Purleemil would result in similar offspring from this wonderful couple, the hunting party rejected the request from Tirtha's tribe to return Purleemil to Tirtha. Instead, they gave Tirtha the opportunity to fight Wimbaco for Purleemil's hand. Cowardly Tirtha refused their offer and left the couple alone.

Negative condition

- pain

- deep hurt

- sadness

Positive outcome

- letting go

- diffuses sad memories

- motivates and re-energises

The following year brought much happiness to Wimbaco and Purleemil including the birth of a handsome son who was the pride and joy of the tribe. Wimbaco would hunt and make toys for his new son, who was destined to become the leader of the tribe, while Purleemil, whose song was renowned, now channelled through the song spirits who taught her even more beautiful songs. Their songs were usually about her little son who was to live eternally and be known as the most beautiful being on the plains of the back country.

The next year the tribe returned to the lake which was close to Tirtha's tribe. More songs came from Purleemil, who said that the spirits were telling her misfortune was at hand. Proud, fearless Wimbaco would not listen to her pleas to abandon the area. As the days passed happily in the new camp, her fears receded and were forgotten and the warning of the spirits ceased.

But one night while the tribe was asleep, Tirtha and his men, who had been patiently awaiting their chance, sneaked up and attacked them. The whole tribe was killed. Tirtha took delight in impaling both Purleemil and her son with his spear. He and his war party left their victims lying on the ground.

The following year they returned and found that the lake was dry and full of salt. The men who had accompanied Tirtha fled in fear. Tirtha stayed on, wanting to gloat over the bones of his enemies, but instead of bones he found masses of brilliant red flowers covering the scene of the massacre. He had never before seen such beautiful flowers. As he gazed, a spear from the sky caught and lifted him from his seat and a voice bellowed, blaming him for his evil actions and for the blood he had spilled, the blood of Wimbaco and Purleemil and their son, which had flowed in one stream and blossomed into the red flowers.

"Their blood shall live forever," he was told. "It shall bloom forever along the bare plains of the salt lakes, which are the dried tears of the song spirits."

The voice told Tirtha that he would sit there forever to pay for his cowardly work. Then the spear impaled him to the ground.

The beautiful red flower—the glory of the western plains—is today called the Sturt Desert Pea. But it was known to the old tribes as the flower of blood. Even today the Sturt Desert Pea is a symbol for many Aboriginals of their ancestors' blood that has been spilled during the two hundred years of white colonisation.

A case history that shows the strength and power of the plant comes from a patient who presented with extreme pain caused by rheumatoid arthritis, with swelling in the joints and pain on resting or movement. These symptoms had been developing for many years, ever since her separation from her husband. He had left her for another woman and she had been very hurt by this. The pain had started very soon after. She had tried a vast number of treatments—herbal, homoeopathic, nutritional, visualisations, affirmations—without much effect. By the time I saw her she could hardly walk or sit for any length of time.

I prescribed Sturt Desert Pea, and it was exciting when she called two days later to say that for the first time in months she had experienced absolutely no pain. She had woken up free from any pain, the swelling had been reduced and she had also released a great deal of sadness which she had been holding inside ever since the separation. The relief obtained was permanent.

The Sturt Desert Pea essence helps to diffuse sad memories after long, old hurts and sorrows. A question that I ask all my patients is whether there has ever been an event in their lives since which they haven't felt well. So many people are holding on to emotional pain caused by the loss of a loved one or a separation. Men, for example, rarely get over their first true love easily, probably because many don't cry to release the pain. They often hold on to the hurt and sadness from a past love affair.

Some of my patients ask me to prescribe Sturt Desert Pea because they have seen such powerful results in other members of their families. Anyone who feels really drawn to a remedy invariably needs it. Many health practitioners who use electronic diagnostic machines such as Vega or Morey machines find that Sturt Desert Pea will clear a multitude of illnesses, ranging from viruses through to parasitic and bacterial conditions. Sturt Desert Pea is a very deep-acting and powerful remedy which helps to bring about profound changes in people's lives.

I now release the pain in my past.
I am now able to express my feelings of sadness and grief.

STURT DESERT ROSE

(Gossypium sturtianum)

Never feel guilty about learning.
Never feel guilty about wisdom.
That is called enlightenment.
            —Ramtha, Ramtha

Sturt Desert Rose grows on the stony or rocky slopes of dry creek beds throughout much of arid Australia—the southern parts of the Northern Territory, north-eastern South Australia, western Queensland, western New South Wales and parts of northern Western Australia.

It is a compact shrub between 1 and 2 metres tall with dark green leaves 5 centimetres long. Its hibiscus-like flower is mauve with a dark crimson centre and has five petals and five sepals. Both the petals and the leaves have black dots. Its numerous stamens unite at the base to form a close-fitting column around the style. The plant usually flowers in winter and the flowers close at night.

The species name honours Captain Charles Sturt, the inland explorer who first discovered this plant in 1884. It was proclaimed in 1961 as the floral emblem of the Northern Territory.

The Sturt Desert Rose essence helps people follow their own inner convictions and morality and do what they know they have to do. Sturt Desert Rose gives people the strength to be true to themselves.

It can be very difficult for individuals to do what they have to do if this is contrary to group morality or pressure. An example is the tremendous peer pressure at parties where drugs or alcohol are being consumed and where one person does not want to participate. Teenagers in particular have a very strong need to be accepted and consequently they find it hard not to conform.

This remedy will also restore a person's self-esteem. Five Corners is for low self-esteem, but Sturt Desert Rose is for a lack of self-esteem that is a consequence of a person's own past action or actions—actions he or she feels bad about. Guilt has a great effect on self-esteem. Sturt Desert Rose is a wonderful remedy for guilt, which can be in the form of regrets about past actions or past inaction. Regrets can hold people back and can prevent the most wonderful things taking place in their lives. Guilt can also lead to self-criticism and fault-finding and produces a general heaviness in one's being. As one young parent found after the death of her child: "After having taken the remedy for only a few days, I was able to sense a release of the great heaviness I had previously experienced in many aspects of my life." Guilt can stem from times when people haven't followed their own morality or from situations that have been left incomplete.

Sometimes guilt is of a less conscious nature. For example, one man took this remedy after talking to his wife about the guilt he felt about his first marriage and the situation that had arisen with his daughter from that marriage. On the second night of taking the bush essence he had a dream where a business associate and friend came to him and told him it was okay and not to worry. That same business

Negative condition

•

guilt

•

low self-esteem

•

easily led

Positive outcome

•

courage

•

conviction

•

true to self

•

integrity

associate had died of a heart attack caused partly by the stress of the venture that they had both been involved in. Although this man did not think that his friend's death had affected him, guilt was obviously below the surface as a few days after the dream he experienced feelings of great relief and elation and ever since he has gained more confidence.

Of the many emotional blocks, guilt probably has the most obvious origins. Young children are told that they are bad by their parents and that they shouldn't burp, shouldn't fart, shouldn't cry, shouldn't touch their genitals. All this responsibility is placed on tiny little toddlers. Being told that by having a tantrum they have totally ruined their parents' day is such a burden for these young souls to carry.

Sturt Desert Rose types can be very apologetic. With some of them, every sentence contains an apology. It is interesting to listen to the language people use as it reflects their attitudes and beliefs. For example, someone who needs Sturt Desert Rose many say, "Oh, I am *so* sorry. Aren't I stupid?" There is great power in words, and if your statements infer something negative about yourself, you will end up creating this condition in yourself. Moreover, some of these people tend to wallow in their own inadequacy as a form of self-indulgence.

They may also be paranoid and feel that others are talking about them or criticising them because that is how they judge themselves. Sturt Desert Rose types may also set such high standards for themselves that they find it difficult to live up to them. Thus they often feel guilty and inadequate. Obligation and duty are very important to these people. In extreme cases, they are irrational and blame themselves when things go wrong for others. If someone close to them is harmed, they believe that they are responsible and that if only they had said or done something different, it would never have happened.

The major cause of sexually transmitted diseases and problems relating to the reproductive system is guilt. In our culture, a tremendous amount of guilt is associated with sexuality. Many children are taught from an early age that it is not right to explore their bodies. As teenagers, they may be told that masturbation is wrong and that they should not experiment with sex. However, while much of our society condemns sex before marriage, advertising promotes its products by associating them with sexually appealing images. These conflicting messages may create a great deal of confusion in the minds of teenagers.

The film *One Flew Over the Cuckoo's Nest* provides a very powerful portrayal of how guilt can be used to manipulate and crush people. One of the asylum inmates, Billy, is beginning to overcome his inhibitions and develop self-esteem through a relationship with a woman, when the matron, a very hard, cruel woman, tries to make him feel ashamed of himself. When this fails, she threatens to tell his mother and infers that she will be most upset. Overcome by guilt, Billy commits suicide.

Religion may also encourage guilt by emphasising that we are all sinners, rather than acknowledging and celebrating the divinity within us all and the joy we can experience through our relationship with God.

When working with people who tend to feel guilty, it is best to ask them to aim for only easily obtainable goals so that they can break their pattern of criticising themselves for what they haven't achieved. Set tasks that will take their attention away from their negativity. This is especially important when dealing with people who are depressed as they need to get moving again so that their life force can be strengthened. It is very hard to remain depressed and guilty when you are active.

The positive aspects of this remedy are a sense of conviction and personal integrity. These people do what they know they have to do and accept what has happened in the past so that they can move on.

I am now releasing all feelings of guilt and regret about the past.
I am now true to myself in all areas of my life.

Sundew

(Drosera spathulata)

He who is outside the door has already a good part of his journey behind him.—Dutch proverb

The beauty of its flat, glittering red rosettes of leaves belies the deadliness of this plant to insects. This insectivorous plant is widespread in swamps, on the banks of creeks and at the bases of cliffs.

The spoon-shaped leaves, which are 3 centimetres long, are covered in fine hairs which exude sticky, dewlike drops that glisten in the early morning sunlight. Insects that are drawn to the plant become ensnared in these hairs merely by touching one, as this stimulates the other tentacles to close over the victim until it is digested. A few days later the tentacles unfold

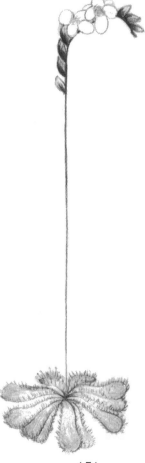

and await more prey. These fine tentacles are incredibly sensitive. The famous proponent of evolution, Charles Darwin, found that it took only a weight of .0000125 of a gram to stimulate them.

The white flowers are only 2 millimetres long and range from pink to white. They arise on a long stem. Flowering time is in spring, although the rosettes remain all year round. Even during summer these plants have the ability to survive the drying out of the soil.

Sundews are found predominantly in South America and in the south-west of Western Australia. *Drosera spathulata*, however, is found in Asia, on many islands of the Pacific, and in New Zealand and Australia, where it occurs along the east coast from North Queensland to Tasmania. It is commonly called the Spoonleaf Sundew.

A very strange thing happened to the medium who checked the information on this remedy. She had a very distinct feeling of dissolving into a blob of liquid and a tremendous urge to be connected to the earth by lying on the floor and covering as big an area as possible. She needed to be grounded, and this is the exact property of the Sundew and also the Red Lily remedy. These people often lack focus. They are vague and indecisive and don't pay attention to detail. They are emotionally split, especially when there is work or something unpleasant to be done—and to them life itself may be unpleasant. Sundew people are often daydreamers, and this remedy will keep them in the present—most of the time.

Using the Doctrine of Signatures, the tiny Sundew flowers are raised high above the solid base of the leaf rosettes and wobble around in the breeze.

This remedy is very good for people who have a broad sense of how something should be done but have difficulty with the fine detail. Sundew helps them focus on the small aspects of a task and it may also help a person to stop procrastinating.

Sundew is one of the four remedies that make up Emergency Essence and is for vagueness or disassociatedness, which is a way of escaping from situations or from life. After a traumatic experience, people often disassociate themselves from the present. At times this is a positive way of coping, for with acute trauma, such as that caused by a car accident, the spirit will leave the body as the situation is too traumatic for the person to experience consciously. When the spirit leaves the body, the astral body is still attached to the physical body by a silver cord. When astral travelling, people can observe their bodies from a distance. This can be done consciously, though it is a dangerous practice, or can be caused by trauma. With a near-death experience, people are drawn towards a beautiful golden light but are told to go back to their bodies because their lives are not over yet. The Sundew

Negative condition

•

vagueness

•

disconnectedness

•

split

•

indecisive

•

lack of focus

•

daydreaming

Positive outcome

•

attention to detail

•

grounded

•

focused

•

living in the present

remedy helps these people return to the physical and the present.

Sundew is also good for a person who is fainting, or recovering from an anaesthetic or in a coma. Of course, the polite form of suicide is just to daydream your life away—not to be present.

On the positive side, when people who are in control of their physical and spiritual bodies, such as Aboriginal elders or other spiritual masters, are at the point when it is time to die and leave their bodies, they are able to do so consciously.

When working with the Sundew remedy, especially for a constitutional condition, one aspect to consider is that vagueness may mask resentment on a deeper level. These people may resent the fact that others are not paying much attention to them, so they withdraw into their own thoughts and ideas and find these far more interesting than what is happening in the outside world. They will often choose to stay split until others show more interest in them and treat them differently.

A number of indications for Sundew may be seen in the iris. There can be gaps in the nerve wreath, indicating that the nerve supply is very disrupted by unclear messages. In some cases the nerve wreath is actually broken, indicating a lack of energy in that area. Excessive amounts of hallucinogenic drugs, even marijuana, can lead to a doubling of the nerve wreath. Even irregular pupil size—one large and one small—can be a sign of past concussion or of being split emotionally.

The positive aspects of Sundew are a lively interest in the outside world and the ability to pay attention to detail and to make decisions. These people are able to channel ideas and information from Spirit and put them to practical use. They are open to inspiration and are able to apply it to the material world in order to benefit themselves and humanity. Inspiration is continually offered only to those who show that they are doing something with it.

As a schoolteacher commented, "Sundew has a grounding effect. It makes you come to terms with reality; it centres your existence and puts it back in your control." Students find it good for improving their concentration, especially if they are drowsy or vague or can't get started on projects.

Having a dream is a Sundew experience, and there is a whole science dealing with how to develop your ability to recall and interpret your dreams. Sundew, Bush Fuchsia, Bush Iris and Isopogon combined will help people not only to dream more often, but to recall their dreams.

Sundew is used with Little Flannel Flower to help children have clearer contact with their spirit guides. Children are very psychic and spend most of their time in the etheric plane when they are first born. They gradually come to integrate the spiritual with the physical. Entities such as gnomes and fairies do exist, and children have the ability to see them and play with them. But children are practical, too, and can balance the spiritual and the physical planes.

I am focused in the here and now.
I am now grounded in the physical plane.

SUNSHINE WATTLE

(Acacia terminalis)

What we are today comes from our thoughts of yesterday and our present thoughts build our life of tomorrow: our life is the creation of our mind.—Buddha

Acacias are the most common and widespread of all Australian plants. Wattle is Australia's unofficial floral emblem. The traditional colours of Australia—blue and gold—depicted the glorious gold of wattle across the blue Australian sky. There are 445 indigenous *Acacia* species. Acacias are also found in Africa, South and Central America and Asia.

The genus derives its name from the thorny Egyptian tree known as *akakia*. The common name wattle is from an Olde English term meaning covering. Wattle stems made ideal reinforcement for the walls and roofs of houses made of mud, hence the term "wattle and daub".

Most wattles are without true leaves, their function being taken over by an expanded leafstalk, an evolutionary adaptation more suitable for the harsh, arid climate of Australia and enabling the plants to survive long droughts.

Sunshine Wattle is an open shrub up to 2 metres tall with pale golden flowers. The fluffy flower heads are actually clusters of between six and fifteen small flowers with long yellow stamens containing large amounts of pollen.

Its habitat is the scrub woodlands and dry sclerophyll forests of the Sydney region, especially on rocky hillsides with poor soil. It is also found in Victoria and Tasmania.

The Sunshine Wattle remedy helps to bring about an acceptance of the beauty and joy of the present and a belief in possibilities for the future. It is usually for people who have had a difficult time in the past and are literally stuck back there. They bring their negative experiences from the past into the present. They have no hope that things will

Negative condition

•

stuck in the past

•

expectation of grim future

•

hopelessness

Positive outcome

•

optimism

•

acceptance of the beauty and joy in the present

•

joyful expectation

improve in the future. What they see can be compared to a mirror that reflects the past, and what they perceive in the present and for the future is only a repetition of what has been.

These people have a strong belief that life is a very grim struggle. They often see only bleakness in life and expect hard times and disappointments again. They are stuck in the unpleasant experiences of the past. This essence helps them see situations from a different point of view.

Sunshine Wattle was used by the First Fleet settlers for their buildings. Life was very harsh in the new colony as they arrived to find that there were no domesticated animals or crops. They had brought little food with them and nearly starved numerous times in those first few years. Unbeknown to them, the supply ship sent six months after they had left England ran aground off the Cape of Good Hope. So after a year, when no new supplies had arrived, they felt totally forgotten and abandoned, and survival was a great struggle. Ironically, the Aboriginals ate the extremely nutritious seeds of Sunshine Wattle, which are exceptionally high in protein. The white settlers must have been intuitively guided to use Sunshine Wattle in their buildings as its essence enables people to see the beauty, joy and excitement of the present and to feel optimism for the future.

It is a good remedy to take when life is temporarily difficult, when nothing is working out or when life seems to be one big struggle.

A good friend of mine moved interstate to take up a position with wonderful opportunities. Previously he had been battling to bring up a young family without much money, which lowered his self-esteem and made him wonder how he was going to make ends meet. When this new challenge arose, he found himself thinking about the struggles of the past and the hardships they had endured and expecting similar situations in the future. Sunshine Wattle helped him move through those feelings very quickly and welcome the opportunities offered by his new circumstances. From this optimistic viewpoint he surmounted some exceptionally difficult problems at the beginning of his employment.

For people who are constantly daydreaming and thinking about happy times in the past—the good old days—Sundew is possibly more appropriate than Sunshine Wattle. Sunshine Wattle people don't regard the past as joyful at all.

A typical Sunshine Wattle story concerns a grandmother who was disinterested in life and for whom everything was an effort. Her grandchildren didn't like to visit her, and her friends and family saw her purely out of a sense of duty, rarely enjoying the time spent with her. After a course of Sunshine Wattle she started having little parties and inviting friends over, buying new clothes, painting the house and playing with her grandchildren again. They began to enjoy visiting her. She developed an interest in people and would lovingly chastise them when they were despondent and try to help them regain their good humour.

This essence has produced excellent results in people who worry about money and who feel that their financial position is becoming worse. Sunshine Wattle has changed their outlook and restored their optimism. Quite often, once they get on with living and enjoying their lives, their money problems have been resolved.

I am now releasing my past and seeing joy and beauty in my life.
I live fully and joyously in the present.
I now feel optimistic about life and the future.

TALL YELLOW TOP

(Senecio magnificus)

One of the most important results you can bring into the world is that you really want to be—Robert Fritz, The Path of Least Resistance

About 1300 species of *Senecio* exist in the world. The name of this genus was derived from the Latin *senex*, which means old man, referring to the white, beardlike scales on the fruit. There are approximately forty Australian species which grow in every State. They are either annuals or perennials, and all are yellow-flowered.

Tall Yellow Top grows up to a metre tall and can be seen on river banks and the dry creek beds of central Australia. It has many small, daisylike flowers packed into each flower head, which looks like a single large flower. Its foliage is soft and delicate. When the winter rains reach the interior, carpets of the massed yellow flowers create a blaze of colour. The

flower is an everlasting, which last century was commonly referred to as an immortelle.

The *Senecio* species belong to the daisy family, which is the third-largest plant family in Australia. This family is also one of the largest and most widespread in the world, with over 20,000 species.

Although this plant has such strong connections with the multitude of plants in its family, the Tall Yellow Top remedy is for isolation and loneliness. This essence addresses alienation, where there is a lack of connection with or no sense of belonging to anything—family, workplace or country. It is also for the lack of love, as Tall Yellow Top helps to reconnect the head and the heart. The colour yellow relates to the intellect. Tall Yellow Top people have such a chasm in their hearts that they are cut off from their feelings and live purely in their heads. These people have usually been in this state for a long time, sometimes even lifetimes. Consequently, they often need to take this essence for a longer period of time, even for six to eight weeks without a break. Its action is like waves pounding on the rocks of the coastline. It has a gradual effect, like the erosion of the rocks by the ocean. Patience is often required when taking this remedy, but its benefits are very profound.

The need for this essence can stem from very early experiences in life, such as abandonment while in the womb or after birth. The parents may not have wanted their child and may have given it up for adoption. Tall Yellow Top is invaluable whenever a person has the feeling of not belonging, of not being wanted or needed, or of lacking a sense of "home". Many people identify very strongly with their work, and being retrenched or asked to leave can result in feelings of alienation and low self-worth.

Many people throughout the world today are homeless. The Palestinians, the Amazon Indians, the Burmese, the Kampucheans and millions of other refugees have lost their homeland. In Australia many Aboriginals are alienated from their own land and culture. They feel they don't belong in their own country. They cannot fit in or are not accepted by Western culture and yet are lost to their own people. Drugs, despair and alcohol often await them as they travel down this road to no-man's-land. For all such cases the Tall Yellow Top essence can be used.

To my mind, the greatest healing event in Australia's recent history occurred on the long weekend in October 1987, the day the Vietnam veterans marched through the streets. There were 45,000 men and women involved in the Vietnam War, and they paid a terrible price for their participation. These people had many diferent motives for going to Vietnam.

Every nation has its warriors, and during the sixties many Australians joined the army as a means of expressing their

Negative condition

•

alienation

•

lonely

•

isolated

Positive outcome

•

sense of belonging

•

acceptance of self and others

•

knowing that one is "home"

— 158 —

nature. In modern times, the warrior is rarely understood by the rest of society. Some felt that they were following in the Anzac tradition, which has always been glorified. Others believed that they needed to protect democracy and Australia from the perceived communist threat. Others were conscripted.

When in Vietnam, it didn't take them very long to realise that the reality was very different from their expectations. The majority of the Vietnamese didn't want them there and didn't care about them. People who sold them food during the day tried to kill them at night. Corruption was everywhere. News about the war in the American and Australian media was totally different from what was really happening. It was a guerilla war in which the Allied troops lived on their nerves. There was more psychic attack and use of dark forces than in any other war. It was a war of mines and snipers, seeing friends blown up by booby traps.

Naturally enough, the army doesn't encourage its soldiers to get in touch with their feelings, because they would fall apart. They experienced intense emotions, and the only way that most of them could cope was to suppress their feelings. When they returned to Australia, they had no outlet for their pain. They sneaked in at four in the morning to avoid the anti-war protesters who turned their anger on the soldiers instead of the politicians. But the soldiers were merely pawns in the war. Employers would not hire them, strangers would come up to them on the streets and ask them what it was like to kill someone.

Their experiences during the war were so different from those of normal civilian life that they were unable to talk about them. Then there were the devastating effects of Agent Orange. Contrary to government claims that it was harmless, their children suffered from birth defects and they experienced continual skin rashes and deteriorating health. Because of all these stresses, their marriages often fell apart. Some of the veterans committed suicide and others were left broken and defeated.

But on the day in 1987 when they marched, the rest of Australia opened its heart and welcomed them back in. The streets were packed with people clapping and cheering them. It was said that every person who marched that day cried, as did most of the others on the streets and many throughout the country who watched the march on the news and read the newspaper reports. The pain and alienation of the Vietnam veterans were washed away by the tears. Tall Yellow Top has also played a powerful role in helping to heal these people.

This remedy works very well for people who feel as though they are "strangers in a strange land", that they don't belong on this planet. They feel as if, through some mistake, they have been sent to the wrong destination, where they just don't fit in. The Tall Yellow Top essence is very effective for those feelings of alienation and brings about a sense of belonging.

I am now able to express the feelings in my heart.
I am now opening my heart and connecting to other people and to Mother Earth.

TURKEY BUSH

(Calytrix exstipulata)

If you 'are seeking creative ideas, go out walking. Angels whisper to a man when he goes for a walk.—Raymond Inmon in A Bag of Jewels *(eds Susan Hayward & Malcolm Cohan)*

Turkey Bush can be found along roadsides and in poor, gravelly soil throughout lowlands and exposed sandstone plateaus at the Top End of Australia. It is an attractive shrub up to three metres high with fissured, twisted stems. The minute leaves, resembling small, green scales, are closely attached to the branchlets, which serves to reduce water loss during the dry season by decreasing the leaf-surface area exposed to the sun. It flowers from May to August with a prolific display of showy, pink-purple, starlike flowers. The petals are narrow and their colour fades as they age.

This remedy came from Katherine Gorge at the Top End of Australia, which consists of thirteen exquisitely beautiful gorges. It is appropriate that the remedy was made up in that area as it contains the world's oldest-known art in the form of Aboriginal rock paintings up to 25,000 years old. Turkey Bush helps people tune into and express their creativity. It can help them when their creative energies are blocked. Emotional blocks and traumas often hold back our creativity. This essence will help to clear such blockages. When adults and teachers analyse children's painting and drawing, they sometimes crush their creativity, especially if they criticise the children's efforts.

For artists, Turkey Bush can help them transcend limitations and keep in touch with their creative inspiration. It is of great assistance to writers, painters and musicians who are going through a period when they lack inspiration. This remedy allows them to contact their Higher Selves and tap into their creativity.

Negative condition

•

creative block

•

disbelief in own creative ability

Positive outcome

•

inspired creativity

•

creative expression

•

focus

•

renews artistic confidence

Many people feel that they cannot express themselves creatively. In our workshops, Kristin holds a section in which participants are encouraged to paint or draw the flowers from which the bush essences are made. Often a dose of Turkey Bush will help them feel confident and comfortable about doing so. The results have been stunning. At one Bush Essence workshop Kristin had all the paintings framed, hired a gallery for a weekend and staged an exhibition complete with an opening night. Those who had thought that they couldn't paint were surprised to find their friends and families admiring their works. We have had similarly wonderful reports from other people who have used Turkey Bush in their workshops. A typical example was given by Anna Gibson, who ran a counselling communication course in Adelaide:

I led a creative drawing and storytelling session where I made Turkey Bush available and explained what it was for. Once students had completed their drawings they gave feedback about the pictures. One of the students said that it was the first picture that he had drawn since being small at school and that he had always avoided drawing and art and couldn't believe that he had drawn something without difficulty!

Turkey Bush provides a means for people to come to terms with their own unique expression. If you stop painting at, say, the age of seven and then start again as an adult, your paintings will be similar to those produced by that age group. For many adults, producing a childlike painting is very disconcerting, yet it is only practice that will bring about a change in style. Turkey Bush can give people confidence in their creativity and the ability to accept and enjoy their own creative expression.

Drawing a plant is a truly wonderful way of getting in tune with it. You can reach a plant's essence by keeping a calm, clear and open mind and just being aware of any perceptions or impressions that you receive. Peaceful feelings, unusual thoughts or even violent reactions may come to you, because in such a state of mind you become aware of influences from another area of the psyche. You are looking at the plant in a different way from your normal perception, and many people confess that they have never looked at anything in this way before.

Tapping creative forces is a very powerful tool which can be a healing force in many subtle ways. Many uterine problems, including infertility, stem from an inability to create, and Turkey Bush has resolved many such conditions. Sometimes Turkey Bush is taken in conjunction with other remedies, especially She Oak, and at other times alone.

I am now tapping into the universal well of creativity.
I am becoming more confident about and enjoying my ability to create.

WARATAH

(Telopea speciosissima)

Faith is the bird that feels the light when the dawn is still dark.—
Rabindranath Tagore

The origins of the *Telopea* genus, one of the oldest plant groups in the world, can be traced back over sixty million years ago to Antarctica. In its more recent history, in fact, in 1962, the splendid Waratah flower was made the official emblem

of the State of New South Wales. It is the only Australian floral emblem known by its Aboriginal name. The word "Waratah" means beautiful. It seems that the people of Australia have always felt this way about the plant. *Telopea* means seen from afar, referring to its majestic appearance which makes it stand out from other flora, while *speciosissima* means most beautiful. The early white settlers described the Waratah as the glory of the Australian bush, as it was the most magnificent plant in the new colony.

Australia does not have an official floral emblem, though Golden Wattle, or *Acacia pycnantha*, is popularly regarded as such. Around the time of the Federation of the Australian States in 1901 there was a remarkable upsurge in nationalism. A great debate raged over the choice of a national floral emblem. The two main contenders were Wattle and Waratah. The Waratah is unique to Australia and is exceptionally beautiful, but it is only found in the eastern States. Wattle is not unique to Australia. Even today, 200 years after the First Fleet arrived, Australia is still without a national floral emblem, and the argument continues. Henry Lawson, the great Australian literary figure, once wrote in a poem: "And I love the Great Land where the Waratah grows." To this day, Australia is still often referred to as the land of the Waratahs.

The *Telopea* genus belongs to the Proteaceae family and consists of four species. The most common form is *Telopea speciosissima*, which is confined to New South Wales.

Negative condition

- black despair

- hopelessness

- inability to respond to crisis

Positive outcome

- courage

- tenacity

- adaptability

- strong faith

- enhancement of survival skills

This Waratah is a stout, erect shrub which grows up to about three or four metres tall. The gorgeous red bloom, which is up to 12 centimetres across, actually consists of many individual flowers packed tightly into a spherical head and surrounded by bright, petal-like bracts. The bracts are very important as they protect and hold the flowers very closely and firmly together. They are like hands holding something precious and frail and convey a nurturing, protective energy. The flower has also been likened to the Sacred Heart of Jesus.

During workshops I often show slides of the flower and ask the participants, who usually know very little about the plant, to describe the impressions they receive when looking at it. They invariably perceive the qualities of strength, courage, tenacity and endurance. These are the very qualities of the essence.

This remedy is for those who are going through a "black night of the soul", those who are in utter despair. It gives them the courage and strength to cope with their crises. This essence is certainly appropriate for those undergoing the ravages of war or the disasters of a stock market collapse. After the Wall Street crash in the 1930s, a popular pastime for business people was to jump from the top of tall buildings. These people felt that losing their wealth was the end of everything.

During times of crisis the Waratah essence will bring to the fore any survival skills that one has previously learnt. It will not only bring forward these old skills but will also amplify them. It is very much a survival remedy which embodies the qualities of the Australian bush dweller—being adaptable and able to cope successfully with all sorts of emergencies. It is from the challenges of life that strength of character is built. Or, as J. Willard Marriott wrote: "Good timber does not grow with ease; the stronger the wind, the stronger the trees."

The Waratah remedy helps to engender courage. If people have some courage it will be increased. Likewise, their skills will be enhanced as they will have the courage to try. In times of crisis, they will be able to call on reserves of strength that they already possess.

Most of the remedies are self-adjusting, but Waratah is not to the same degree. As it is also one of the most powerful of all the essences, it is best used for shorter periods of time. Because of the conditions that this remedy addresses, it needs to be quick-acting. The initial benefits are immediate, and in many cases the full effects are achieved in only a few days. Others get good results within five to seven days. Waratah and Sturt Desert Pea are the two most powerful essences.

While writing this entry on the Waratah, I received a long-distance telephone call from a terribly distressed woman who wanted a bottle of the Waratah essence for her son, who had just attempted suicide. Such is the magic of the synchronicity that is so often observed with the bush essences. Waratah has been of great benefit to people who have experienced deep despair or suicidal feelings. It helps them find their faith and the courage to keep going. We never experience anything in life that we do not have the strength and ability to deal with. Waratah can help to bring us this awareness and understanding. Many lives have already been saved by this single remedy. No wonder so many people have remarked, "It's such a powerful essence!"

Because of the power and beauty of this plant, its form is widely used in art, architecture and design. Many corporations have adopted the flower as their logo, and it has always played a very significant role in Aboriginal culture. The Aboriginals of the Tharawal tribe around the Cronulla region of southern Sydney used the Waratah medicinally. They placed the flowers into a bowl of water, so that the nectar would be soaked out. The water was then drunk for pleasure, for its strengthening effect and for curing illnesses in children and elderly tribe members.

The flower is very common in the folklore and legends of the Aboriginals. The following legend is one of my favourites:

Long ago, there lived a beautiful Aboriginal woman called Krubi. She lived with her tribe in the Burragorang Valley. She was easily recognised, for she had a unique cloak made from the red skins of the rock wallaby. It was ornamented with the red crests of the Gang Gang cockatoo. There was nothing more beautiful in the world than her cloak. Krubi was in love with a young man from the tribe and she would wait for her man to return with the hunting party each day, from the highest point. The first thing he would see was her red cloak. This was the thing he longed he see.

But one night Krubi was deeply upset, for she learned that the men were going out on a war party, as Aboriginals from another tribe were trespassing on their land. The next day she again stood on the sandstone cliff to await her lover. From there she could hear the yells and cries of the battle and was later greeted by the return of ragged fighting men. However, the man of her heart was not with them. She waited up there for seven days hoping for him to come back. During that time her tears formed rivulets in the sandstone. From her tears sprung new plants, the Boronia, the Eriostemon and the Bush Fuchsia. After the seven days she went to the battle scene but could find no trace of her lover. Krubi returned to her sandstone ridge and willed herself to die. As her spirit passed through a crack of sandstone, up came the most beautiful of all Australian plants, with a firm, straight stalk. It was perfect in beauty, just like the man Krubi had died for. The leaves were serrated and pointed, like his spear. The flower was more red and glowing than any other, and just like her red cloak it could be seen from far away.

Legend also has it that the King of the Burragorang tribe presented an early white Governor with a Waratah as a peace offering, as it was the national flower of the Aboriginals. It was said to be the only flower they ever plucked to show or give to white settlers.

A second legend, from *Aboriginal Australia* by Burnum Burnum, describes how the white Waratah became red. (A white Waratah does exist, both in New South Wales and Tasmania.)

Long ago in the Dreamtime, all Waratahs were white. At that time, the first wonga pigeon camped in the bush with her mate and they grew fat on the abundant food to be found at ground level. They never flew above the trees, because they were afraid of their enemy, the hawk.

One day the wonga pigeon's mate went searching for food and failed to return. She searched a long time for him without success and finally resolved to fly above the treetops to see if she could see him from above. As she left the shelter of the trees, she heard her mate call from down in the bush and, with a glad heart, she turned to fly down to him. But the circling hawk had already seen her and swooped down, grasping her in his sharp claws and tearing open her breast. Wrenching herself free from the hawk, she hid among the blossoms of the Waratahs. The hawk could not find her and flew away. Again she heard her mate calling. Weak from loss of blood, she tried desperately to reach him, but she could only fly short distances and every time she rested it was on a white Waratah, her blood staining the blossom and turning it red.

As her life ebbed away, she changed the white Waratahs to red, and today it is rare to find a Waratah that is not tainted with the blood of the brave and loyal wonga pigeon, who lost her life searching for her mate.

We chose the Waratah as our logo because we feel that it is a very important remedy which will become even more important in the years to come. Metaphysically, major spiritual, economic, social and physical changes are common at the end of an age, and the next is due around the year 2000. If those upheavals occur, literally millions of people will be taken totally by surprise, will have no understanding of what is happening and will quite likely find it very difficult to cope. The Waratah essence will help them find their own faith and will give them the courage and strength to deal with those changes. It is a very powerful and important remedy.

I have courage, tenacity and faith in all situations now.
I am now finding the courage and strength to face the challenges of life.

WEDDING BUSH

(Ricinocarpus pinifolius)

Concerning all acts of initiative (and creation), there is one elementary truth, the ignorance of which kills countless ideas and splendid plans; that the moment one definitely commits oneself, then Providence moves too.—Goethe Faust: A Tragedy

This bushy, well-branched shrub grows up to 2 metres high and is commonly found in poor, sandy soils on sandstone plateaus in Queensland and New South Wales. Wedding Bush's flowers are white, like most flowers pollinated by beetles, which have poor eyesight. The shrub has an abundance of these unisexual flowers which occur in groups consisting of a few male flowers and a single female. Each flower has four to six petals which are 10 to 15 millimetres long. At weddings in outback Australia in times gone by, the flowers were worn in bridal veils, hence the plant's common name.

This essence is for people who have difficulty in making commitments, whether to relationships, employment, the family, or to their own personal goals. Such people often appear to be running away from themselves and avoiding responsibility. It is an excellent remedy to take when beginning a partnership or for any form of bonding. It can assist in providing dedication to a task or relationship or even to one's life purpose.

Wedding Bush can be likened to the cement that keeps a relationship together. It is helpful when people want to recommit themselves to each other, or when a person wishes to stop going from one relationship to another. The latter pattern has been referred to as the Casanova Complex, which describes a well observed and documented male response to women and relationships. After the initial physical attraction is gone and a greater closeness develops, usually within a few months, these men end their relationships. Richard Bach's autobiography entitled *Bridge Across Forever* is a good example of the situation addressed by Wedding Bush. Bach was confronted with a choice of continuing in one relationship or maintaining his pattern of having numerous short affairs. His experiences are well described and very typical of many males.

Recently one of my patients, a two-year-old girl, was brought in because she was suffering from stomach upsets. After working with her and using kinesiology, I discovered that Wedding Bush was the major remedy. I was surprised by this, but it turned out that the child was reflecting the imbalances in the family, which is known as surrogating. The child's mother was experiencing great conflict because she was having an affair and was uncertain about whether to remain in the marriage or not. After talking to the mother, I prescribed Wedding Bush for the whole family. The mother became clear about what she wanted, choosing to stay in the marriage, while the child's condition quickly cleared up.

Another woman, who had been quite happy in her marriage for fifteen years, had in the last year become totally infatuated with another man and was unable to stop thinking about him. She had told her husband of her attraction to this man, though she had not had an affair with him. When I saw her she was convinced that she was going to have to sleep with him, though she was also very concerned about the effect this would have on her marriage. After taking Wedding Bush for only a day, she came to see me again. She nonchalantly told me that her thoughts and desires for this man had suddenly left her and she was feeling quite content with her husband once more. She continued to take Wedding Bush, but her feelings for this other man remained resolved.

Wedding Bush is very good for people who have taken on activities or regimes that require discipline and commitment.

Negative condition

- difficulty with commitment to relationships

Positive outcome

- commitment to relationships

- dedication to a goal or life purpose

It has proven to be of great benefit to athletes who need to be committed to their training schedules and even to people who wish to become slimmer and to follow a healthier lifestyle.

I made this remedy up on the anniversary of my own wedding. Kristin and I often give it as a gift to friends who are getting married, so that they can take the essence before exchanging wedding vows and then on rising and retiring for the first few days of their marriage.

I am now able to commit myself to whatever I choose to undertake.
I am now dedicated to the fulfilment of my life's purpose.

WILD POTATO BUSH

(Solanum quadriloculatum)

Life is constantly providing us with new funds, new resources, even when we are reduced to immobility.
In life's ledger there is no such thing as frozen assets.—
Henry Miller

There are over eighty species of this genus in Australia, which belongs to the same family as the tomato, the potato and tobacco.

Wild Potato Bush is a perennial shrub with numerous prickles on its leaves and stems. In spring the purplish blue flowers, with their clustered yellow stamens in the centre, are very prominent. The plant is found in various habitats in central Australia and can be frequently seen along roadsides. The tomato-like fruit is not edible, whereas many of the native *Solanum* species make up an important part of the Aboriginal diet.

This plant, like many of the *Solanum* species, has a very unworldly look about it. The remedy is for people who feel burdened by the physical body and restricted by it. They feel the need to step beyond their physical limitations as their bodies are holding them back and weighing them down. They are unable to bring about the changes needed in order to move forward. This essence addresses the frustration of restriction.

This remedy is appropriate for paraplegics, quadraplegics, and those who are overweight or who have an illness that restricts the body. It brings about a sense of vitality and

Negative condition

•

weighed down

•

physically encumbered

•

frustrated

Positive outcome

•

ability to move on in life

•

freedom

•

vitality

— 169 —

freedom. Exercising is a good way to lose weight, but people who are carrying too much weight often have swollen ankles, which restricts their activities. Thus overweight becomes a vicious circle.

Wild Potato Bush is commonly taken by women towards the end of pregnancy, when their movements are restricted. They often find lying down uncomfortable and have to get up frequently at night to urinate. They may also suffer from indigestion. Carrying an extra load can certainly take its toll on a pregnant woman's energy levels.

Yet it is not only the woman who benefits from this remedy. When I use kinesiology to surrogate test the unborn child, the most common bush essence needed is Wild Potato Bush. I have often wondered whether the incarnating soul is having difficulty with being in a limiting physical body once more, or whether the baby feels frustrated because it lacks control over its body. Of course, for the lack of physical control, a baby can be given this remedy at many stages during the first few years of its life. Many parents and health practitioners have noticed that the time of greatest frustration for a baby occurs just prior to taking its first steps. The baby feels ready to walk but its body has not developed sufficiently to hold it steady. It is debatable whether or not a child walks earlier after taking this essence, though many claim that this is so. However, the child's frustration is certainly eased.

After taking Wild Potato Bush, one of my friends had an insight which was later confirmed during age regression work with me. She realised that the reason she had always been overweight and had rarely been able to diet successfully was that the weight served as a form of protection for her. Mary had experienced a difficult birth and was actually stuck while being born. Her head emerged easily but not her shoulders. Consequently, she formed the beliefs that the world isn't a safe place and that she needs as much padding as possible to protect herself. Not surprisingly, she had an exceptionally thin neck in proportion to the rest of her body. She also found that, as her attitude towards her body became more positive and she felt less burdened by it, she began to lose weight.

My body is now full of energy and vitality.
I am releasing all restrictions on my ability to move forward.

WISTERIA

(Wisteria sinensis)

The joy of sexual love, manifested, in so many attractive and delightful ways, makes the human condition truly blessed.—Smaradipika in Sexual Secrets *(Nick Douglas & Penny Slinger)*

Wisteria sinensis, one species in a small genus of attractive, robust, woody climbers with twining stems, is native to North America and eastern Asia. It was introduced into Australia and Europe in the early part of the nineteenth century.

Wisteria is a deciduous vine which can grow to a height of over 30 metres. It ascends by twisting its stem around any suitable support, though it can be trained to form a standard or large bush. It has large, dark green leaves and 30 centimetre-long, drooping clusters of fragrant, mauve, pea-shaped flowers. It requires a temperate, frost-free area in order to thrive.

I certainly remember the day this remedy was made up. It was a wonderful example of the synchronicity associated with the bush essences. Later that day seven of the eight patients I saw were women presenting with sexual problems. Five of them, in fact, were new patients who had booked their apointments weeks earlier.

This remedy relates predominantly to women, but it can also be used by men who have adopted a macho image and deny their feminine aspects. It helps them accept that they have a gentler, softer side, though in order to express this

Negative condition

•

frigidity

•

sexual hysteria

•

"macho male"

Positive outcome

•

sexual enjoyment

•

openness

•

gentleness

—171—

side they may need to take Flannel Flower, possibly for some time!

This remedy is for women who feel generally uncomfortable and uptight about their sexuality. It is helpful for those who feel tense about sex and are unable to relax and enjoy it, or who are afraid of intimacy. In many instances, these feelings are directly attributable to previous sexual abuse or assault. It has been estimated that up to 70 per cent of women are sexually abused or assaulted at some point in their lives. Wisteria, when used in conjunction with Fringed Violet, will help to clear the emotional and physical scarring caused by such events.

A typical Wisteria situation was presented to me by a woman who grew up in a small Victorian country town in the early 1940s. At the ripe old age of twenty-one, with all her girlfriends already married, she imagined she was going to be left on the shelf. She had only ever had one boyfriend, so when he proposed she accepted, not because she loved him but because she thought that if she didn't marry him, no one else would ever ask her.

Her experience of romance and sexuality was based on watching Doris Day and Rock Hudson movies at Saturday matinees. She got a rude shock when confronted with the reality of sex. She didn't enjoy it and switched off in response to her husband's rather rough and insensitive lovemaking.

For the sake of her three children, she stayed in the marriage for a number of years. Eventually, and with a great sense of liberation, she left her husband and later met a very caring and gentle man whom she married. Yet even though she felt great love and tenderness for this man, she found that she still switched off whenever they were intimate. Wisteria taken over a period of time helped to resolve her problem and to create a fulfilling sexual relationship for the couple.

A woman in her late twenties, whom I had already been treating for some time, announced that a Pap smear had revealed cancer cells in her cervix and that her gynaecologist wanted to operate on her. While discussing her sexuality in order to discover why she had created the illness, she informed me that she had never felt comfortable about being touched ever since the age of fourteen, when she was sexually abused by her boyfriend. For this trauma and its consequences, I prescribed the Wisteria–Fringed Violet combination.

When she came back for her next appointment, she mentioned that, after two days of taking the remedy, she had felt intense burning sensations in her vagina. Though this distressed her, she felt that a healing crisis was taking place and continued to take the essence. When she visited the gynaecologist prior to surgery, he was astonished to find that the cancer had vanished.

Again, I must emphasise that the remedy was not used to treat cancer but to treat the emotional imbalance, in keeping with the philosophy that physical illness is only the manifestation of emotional imbalances. Problems with the reproductive area usually reflect emotional problems concerning sexuality.

The reason that some women do not feel relaxed sexually cannot always be explained by instances of rape or assault, as often the cause lies in beliefs developed by the woman while in the womb or in the early years of her life due to her parents' attitude towards sex. Wisteria works very powerfully to resolve those negative beliefs and helps to bring about an enjoyment of sex and an ease with sexual intimacy.

I am now comfortable with and enjoy my sexuality.
I am now open to intimacy in my relationships.
I now realise that I have a gentler, softer side to my nature.

Yellow Cowslip Orchid

(Caladenia flava)

*T*he most effective way to achieve right relations with any living thing
is to look for the best in it, and then help that best into the fullest
expression.—J. Allen Boone, Kinship with All Life

The orchid family contains between 15,000 and 30,000 species,
more than any other plant family. The beauty and
ornamentation of orchids are rarely matched by other plants.
Orchids are considered to be the most highly evolved form
of plant life, due to both their floral structure and the highly
complex, ingenious way in which they attract insects to
pollinate their flowers.

Australia has over 600 orchid species. The *Caladenia* genus
consists of some of the most attractive of all the ground
orchids. *Caladenia flava* is a native of Western Australia found

from Kalbarri in the north to Esperance in the south-east. It is a widespread species which prefers light, well-drained soils in shady, protected spots. It has broad, hairy, lance-shaped leaves, while its yellow flowers have red-striped petals and are 3 centimetres across. The plant grows to a height of up to 15 centimetres with a long, thin, erect stem.

Yellow is the colour symbolising the element of Air which deals with the intellect. The pituitary is the endocrine gland associated with Air and is balanced by the Yellow Cowslip Orchid bush essence.

This orchid has a very social and gregarious nature and is commonly found growing in a cluster. This is another aspect that relates to the Air element, for Air is very closely connected with social order, group activity and harmony—the ordered society. Yellow Cowslip people are focused so much in the intellect that they are often blocked off from many of their feelings. When they are out of balance they have a tendency to be excessively critical and judgmental, as well as aloof, withdrawn and overly cautious about accepting things. They can be very petty, nitpicking and sceptical. Ironically, these are the very qualities that can lead to poor functioning of the pituitary gland, which in turn can result in loss of memory and a weakening of the intellect. If this occurs, the bush essence Isopogon, another yellow flower that relates to the mind, is often indicated, especially when used alternately with Yellow Cowslip Orchid which will deal with the cause of the problem. Women who have been on the Pill for many years and suffering from hormonal imbalances often respond very well when given this essence.

One of the main functions of the pituitary gland is to control the rate of growth. Not surprisingly, Air people are the tallest of all the element types, usually being very thin and having long legs and oval-shaped faces. The pituitary, if stimulated by excessive judgement or criticism, can restart growth of certain anatomical features, such as the nose, in middle or later life. So maybe Pinocchio's problem was not so much telling lies as being too judgmental or critical.

I had been told in meditations many years ago that I would work with fifty bush essences. The discovery of the fiftieth took nearly a year to come about. In fact, I had made up this remedy many months earlier at a workshop in Perth but had not explored its properties to any great extent, though I realised that it had an effect on the pituitary. Months later in Sydney, a woman presented with her young, mentally retarded daughter, saying that a transmedium has given a health reading in which it was claimed that if her daughter used bush essences for about two years, great healing would occur and her intellect would become very close to normal. During the first consultation I realised that the pituitary was her key endocrine gland and the major source of her imbalance.

Negative condition

- critical

- judgmental

- bureaucratic

- nit picking

Positive outcome

- humanitarian concern

- impartial—can step back from emotions

- constructive

- ability to arbitrate

I knew intuitively that Yellow Cowslip Orchid was the remedy that would initially be of most assistance to her. This led me to delve deeper into its healing properties, particularly on the emotional level.

Yellow Cowslip Orchid's positive qualities are an open and inquisitive mind, the ability to grasp concepts quickly, the uncritical acceptance of people and ideas, and the ability to arbitrate fairly and compassionately while considering all of the facts.

I am now releasing all judgement and criticism.
I am now able to see the forest and the trees.

How to
Prepare and Take
—Bush Essences—

Preparation Methods

Use flowers that have grown in an environment away from pollution and power lines. Sensitivity and a degree of reverence are necessary. Place the flowers without directly touching them in a bowl of water and leave this in the sunlight for several hours. Remove the flowers from the bowl, preferably with a twig or a leaf from the same plant. This mother essence water is preserved with an equal amount of Australian brandy. Seven drops of this mother tincture are added to a 15-millilitre bottle containing two-thirds brandy and one-third purified water, to form the concentrate, or stock bottle.

Stock Bottle Use

Dosage bottles are prepared by taking a 15-millilitre dropper bottle filled with three-quarters purified water and one-quarter brandy as a preservative and adding seven drops from the stock bottle. Several essences can be combined in the one bottle, but it is generally suggested that the number combined be limited to four or five.

After the stock has been added, the dosage bottle can be lightly shaken or tapped to release its energy. Some people also energise their essences with a prayer or invocation, or with a visualisation such as surrounding the essence with white or gold light, though none of these is essential. Affirmations can also be used very effectively with the essences.

Dosage Bottle Use

Flower essences are taken from dosage bottles morning and night by putting seven drops under the tongue or by adding the seven drops to water and are usually taken for a two-week period. They can also be applied topically by adding them to lotions and salves, or to bath water (a very effective method).

Choosing and Administering the Essences

Often the excitement or amazement people feel on initially reading about the properties of the bush essences is quickly overtaken by their absolute conviction that they need to take at least half the remedies instantly, because they identify with them so strongly.

This is a very common and normal response! It demonstrates one of the beauties of the essences—namely, their simplicity. Not only can you see yourself in the remedies, but also your auntie, cousins, best friend, nextdoor neighbour and your parents! All this instant insight is available to you without years of formal training, but merely with a basic understanding of human nature.

There are many ways to choose the most appropriate bush essence or combination of essences. You can simply read the descriptions of the individual essences and select the one most relevant to your situation or personality; you can choose the flower whose appearance you feel most drawn to; you can use kinesiology to pinpoint the right essence; or you can select a remedy based on your numerological chart and year cycle. A question to ask yourself that will help you choose an appropriate remedy is, what do I most want in my life? The answer to this question will lead you to the essence that is most relevant to your life at that time.

So, now you have chosen a few remedies that seem appropriate, what's the next step? Well, you have the option of taking either a single essence remedy or a number of essences combined. Generally the effect is finer, more powerful, faster acting and longer lasting if you take an individual remedy. The mixing of two or more unrelated essences can produce more of a physical action in the body. Therefore, you may decide to narrow your choice to the one remedy that most completely addresses the issue, situation or emotion you wish to deal with.

However, certain combinations work exceptionally well—in fact, just as well as a single essence, if the essences in a combination all address different aspects of a particular condition or situation. Some of these combinations are as follows.

Radiation Essence

This combination is made up of Bush Fuchsia, Crowea, Fringed Violet, Mulla Mulla, Paw Paw and Waratah.

The Radiation Essence can be used to negate or reduce earth radiation found where lay lines cross or in houses under which underground streams run; electrical radiation emitted by meter boxes, overhead power lines, fluorescent lights and electrical equipment, especially televisions; solar radiation, such as bad sunburn; and radiation therapy used to treat cancer.

With radiation therapy, this combination helps the normal, healthy cells withstand the radiation and recover after therapy. It would also be useful against nuclear radiation. In all these situations, the combination stops the storage of radiation in the body and helps to emit the radiation already stored, to keep the body's energies intact and the neurological systems functioning normally.

Emergency Essence

The Emergency Essence is a combination of Fringed Violet, Grey Spider Flower, Sundew and Waratah. It has a calming effect on the mind, body and emotions during minor and major crises. It will quickly ease fear, panic, severe mental or physical stress, nervous tension and pain.

If a person needs specialised medical help, this essence will provide comfort until treatment is available. The wide variety of uses for this combination ranges from pre-examination nerves to gross physical injury. Administer this remedy every hour or more frequently if necessary, until the person feels better. It can be mixed into a cream.

Superlearning Essence

This powerful combination of essences is unsurpassed for bringing about mental clarity and focus and enhancing all learning skills and abilities. It comprises Bush Fuchsia, Isopogon, Paw Paw and Sundew.

Personal Power Essence

This combination of Dog Rose, Five Corners, Southern Cross and Sturt Desert Rose brings out our true inherent positive qualities of self-esteem and confidence. It allows us to feel comfortable around other people and resolves negative subconscious beliefs we may hold about ourselves as well as any guilt we may harbour from past actions. This combination also helps us take full responsibility for the situations and events that occur in our lives and realise that we have the ability and power not only to change those events, but also to create those that we want.

Vitality Essence

This combination brings about abundant energy, vitality, enthusiasm and joy for life. This is achieved by balancing and stimulating the major glands associated with energy—the thyroid with Old Man Banksia and the adrenals with Macrocarpa. Also in this essence are Crowea, which balances the organs and muscles, and finally *Banksia robur*, for temporary loss of drive and enthusiasm.

There are many other possible combinations of the bush essences. Experiment with these combinations. Feedback and case histories are invaluable for providing greater understanding of the scope of the bush essences. The Australian Bush Flower Remedy Society regularly publishes case histories in its newsletter. See page 191 for further information on the Society.

Administering the Essences

With Australian Bush Flower Essences, the dose is seven drops on rising and retiring for a period of two weeks. The seven drops can be placed directly under the tongue from the dropper or can be added to a little water, which is sipped. Placing the drops under the tongue enhances absorption. Try to avoid touching the dropper with your mouth, as this tends to create bacterial growth in the essence. When taking the drops in water, hold each sip in the mouth for about ten seconds or so before swallowing.

You can leave the remedy by your bed, and as you get up in the morning and go to bed at night you'll be reminded to take it. It's important to develop the habit of taking the remedy regularly in order to maximise its potential. The times of rising and retiring are key periods for a person's psyche, for then the mind is often in a relaxed and receptive state and the subconscious is easily accessed. It can be beneficial, as you take the remedy, to focus on the positive aspects of the essence and have a sense of, or visualise yourself, achieving these positive results.

After two weeks stop taking the essence and review your position. The emotional state or other issue that you were working on may have been resolved, in which case you can consider whether or not there is another problem to work through.

Frequently a deeper emotion will surface once the top layer of emotion has been peeled away. If this happens, you can then choose the particular essence that best deals with the new situation. Again, take seven drops of the new essence morning and night for two weeks.

Alternatively, having resolved an original issue, you may feel you've regained your "balance" and have no immediate need for more remedies. Remember, though, that the essence will still be acting in your being for some time after you have finished taking it. However, if you feel that the original emotion or issue hasn't been fully resolved, then continue taking it at the same dosage for another two weeks.

Two weeks is usually sufficient time for a bush essence to take effect. However, if an emotion or belief has been running your life for a long period of time, it may take slightly longer for the essence to help you resolve the issue. Among the fifty essences, nearly everyone should be able to find one—a constitutional remedy—that matches their personality more closely than the other essences. The Black-eyed Susan type, for example, is always rushing, striving and on the go and generally has little patience. Now these people have probably been like this all their lives, and obviously Black-eyed Susan taken for two weeks won't bring about a complete, instant and permanent change. Yet the remedy will modify the individual to a point of balance. The remedy may need to be repeated, but perhaps not for another six months or more, and an even longer period is likely to have elapsed, possibly even a year, before it is needed again.

One category of essences can be called the acute remedies. These are the remedies that work straight away. If someone is feeling suicidal, then there's no point in saying, "take this Waratah essence for two weeks and you'll feel better," as they may not be around in two weeks! The acute remedies often don't need to be taken for a full two weeks, as a few days is usually sufficient. Simply stop taking them when the desired result has been achieved. Waratah, Bottlebrush, Bush Fuchsia, Fringed Violet, Grey Spider Flower and Paw Paw are some of the more common acute remedies, though nearly all of the remedies can be used in this way at one time or another.

In the case of a person who is not able to take the drops by mouth—the mouth may be painful, or the person may be unconscious, in which case Emergency Essence may be necessary—rub the drops into the temples, lips or wrists. This is a very effective method of administering the remedies.

For topical application, you can soak in a bath that contains seven drops of the essence and absorb the energy of the remedy this way. This feels quite exquisite! Another method of external application is to put seven drops of the essence into a bowl of purified water—but not distilled water, unless it's been left out in the sun for a day or so. Splash this on to the part of the body to be treated and allow it to dry. You can even spray it on in a fine mist, particularly when using Spinifex for blistering types of skin eruptions and fine cuts, or Billy Goat Plum for skin lesions. Do this morning and night, or more frequently with Emergency Essence for acute trauma. The latter can also be applied to the skin directly in either dosage or stock strength for emergencies such as burns, insect stings, animal bites, falls, etc.

—Sexuality—

The best way to gain an appreciation of the properties of specific essences and their relevance to our lives is to explore a common human theme and to note the essences that are appropriate at various points of that exploration.

In this examination of how the bush essences can be used in relation to sexuality, we begin at the time of conception and follow the person through the usual biological and social landmarks of life.

An individual's sexuality begins to be shaped long before birth. The experiences observed by many rebirthers and therapists have confirmed the view that the emotions present at the time of conception leave a deep impression on the psyche of the child. If a child senses an indifference or coldness between its parents, an insecure attitude towards itself and its sexuality may be created in the child. There is a great difference in the emotional stability of children who are conceived in an act of loving passion and those conceived violently, such as during rape or a drunken rage. Terror or love, whatever is experienced, will be the basic emotion that colours the child's attitude towards the world.

A lack of intimacy between the parents during pregnancy can also influence a child's view of sexuality. Parents who maintain a happy, loving, intimate relationship are likely to have a child with a relaxed and healthy attitude towards sex. Bush Gardenia and Flannel Flower help to address a lack of interest in one's partner and an inability to get close.

Some new mothers feel very unattractive, perhaps because they have put on weight or because their figures have changed (Billy Goat Plum). Their children may pick up these feelings and feel great guilt, as they believe that they are responsible. Also on this theme, an interesting pattern found more commonly in males has been uncovered by rebirthing. Some babies are aware of the pain that their mothers go through during labour and birth and feel responsible for it. This leads to the subconscious belief that having sex is painful for women. Later, when these men have sexual relationships they may suffer from premature ejaculation, as their subconscious belief makes them want to complete the sexual act as quickly as possible. The Sturt Desert Rose remedy is very effective for resolving feelings of guilt.

Many young children enjoy touching their genitals. If these children are forbidden to do so, or are smacked and told that they are dirty, they will not only develop guilt but also a sense that there is something wrong with their bodies—perhaps they will even feel disgusted by the genital area. Again, Sturt Desert Rose will help to remove the guilt, and Billy Goat Plum the feelings of disgust for their bodies.

When the initial sexual attraction towards their partners begins to fade, some men spend a lot of time with their male friends. In order to relieve their loneliness, their wives may shower their children with love. Many of these children, especially the boys, feel smothered by their mothers' constant attentions. As these young men grow up, whenever they are in a relationship with a woman their belief that they will be smothered if they get too close comes up. So if the relationship becomes too intimate, they will withdraw, never really understanding why. Dog Rose can be very effective for the general fear, Wedding Bush for the lack of commitment and Flannel Flower for the inability to get close.

A child's attitude towards the opposite sex is very powerfully influenced by parents' comments and attitudes. If a father constantly denigrates women, then his children's view of women will tend to be negative. Similarly, racist remarks from parents will tend to influence their children's attitudes towards others, perhaps preventing them from forming relationships with people from other races. Slender Rice Flower is the remedy for developing tolerance and an understanding that all people are equal. Cowslip Orchid is for those who are critical and judgmental.

Many children have a poor relationship with their fathers. As a result, they may also have difficulty in relating to other authority figures. As women tend to choose partners who resemble their fathers, those who have a poor relationship with their fathers will tend to experience a lot of friction in their marriages. Red Helmet Orchid can be used to bond fathers with their children, and Bottlebrush is for bonding women with their children.

Sometimes people who have been adopted find it very difficult to form close ties with a partner—to surrender themselves to a relationship. Tall Yellow Top can be used by these people, possibly interspersed with Wedding Bush for helping them make a commitment.

Wedding Bush will also help to address the pattern common among men of not forming any permanent relationship. As soon as the initial sexual attraction starts to fade, they are on the lookout for new conquests. This remedy can, of course, apply to women too.

Puberty brings about many changes and many stresses, too. For a girl, her menarche heralds the transition from girlhood to womanhood, while a boy has to come to terms with his voice breaking and other physical changes. Both have to cope with the rapid physical and emotional demands of their new sexuality, as well as others' reactions to them as sexual beings. Bottlebrush is very helpful for those going through the major biological changes of puberty. Billy Goat Plum can also be used to bring about an acceptance and enjoyment of the body during this period. Sturt Desert Rose can help if guilt arises from the exploration of the body or from masturbation.

Another of the physical changes that can occur at this time is the onset of acne—the scourge of teenagers and sometimes of later life. Again, Billy Goat Plum can be taken for the feelings of self-disgust caused by an outbreak of acne. As acne is usually an outer expression of low self-esteem, Five Corners can also be useful for this condition. Acne often becomes a vicious circle—the worse one feels about oneself, the worse the acne becomes.

A lack of confidence can prevent people from establishing relationships, or it can result in their allowing others to use them sexually. These people often lack enough self-esteem to stand up for themselves and to ask for their needs to be met. Sturt Desert Rose is not only for guilt but also for doing what we know we should do, for following our own morality. The film *Puberty Blues* dealt superbly with adolescent sexuality and accurately portrayed the peer pressure that can so destructively undermine the inner convictions of teenagers.

Courting may also end in the pain of separation. Numerous research papers have shown that, for many men, it takes years to overcome their grief over losing their first love. Sturt Desert Pea is very good if they are still carrying that deep sadness within them. After the break-up of a relationship, Dagger Hakea can help people deal with the resentment they feel towards the other person. The combination of Bottlebrush and Boronia is excellent for helping them let go of the other person and for healing their broken hearts.

Jealousy can also be a problem in relationships. Mountain Devil will help jealous people cope with the intense feelings that sweep over them. Fringed Violet may have its part to play if the jealousy is combined with shock; for example, seeing one's lover out with another person.

Young, inexperienced people can find themselves in relationships that they don't know how to end. They don't want to hurt the other person, yet they feel trapped. When you are in a situation that you feel you can't get out of, think of Red Grevillea.

Kangaroo Paw is very effective for those who have difficulty in relating to the opposite sex.

The loss of virginity may produce many strong emotions. Becoming sexually active, especially for the first time, can trigger fears of pregnancy, AIDS and venereal disease. The remedy in this situation is Dog Rose for general fears, or Grey Spider Flower for terror. For those afraid that their parents may find out, Dog Rose can be taken.

Crowea is the remedy for worry—worrying about whether you are doing the right thing, whether you are kissing in the right way, and worrying about what others think of you.

Flannel Flower and Wisteria are essences that can help men and women, respectively, to enjoy sex and closeness. Billy Goat Plum will allow them to experience the beauty of the sexual act and their bodies, especially if they feel unclean after sex.

A lot of adolescents find dating and the possibility of being rejected very stressful. The Illawarra Flame Tree remedy can be of great benefit when dealing with these feelings.

People who experience the same problems in one relationship after another will find that Isopogon helps them learn from their experiences, so that they don't have to continue to create similar situations.

After a series of disappointing relationships that have ended early or have not been satisfying, or when there have been no relationships at all for some time, some people give up hope of ever finding a partner. Although they may desperately want a good relationship, these people resign themselves to remaining alone. Sunshine Wattle is a good remedy for these people.

Not only is AIDS a very great threat today, but herpes, genital warts, NSU, trichomonas and a whole array of other sexually transmitted diseases are also prevalent. A common reaction from people who have contracted a less serious type of STD is revulsion and disgust. Billy Goat Plum will assist these people to work through their feelings and will encourage them to accept and enjoy their bodies again. Most sexually transmitted diseases are the result of deep feelings of guilt.

For herpes, it has been found that the topical application of Spinifex is very effective in helping to heal the lesions, and when taken internally it will help to bring to the surface the emotional beliefs or patterns that have led to the herpes manifestation. Our physical reality is created solely by our thoughts and unconscious beliefs.

For traumatic sexual experiences such as rape or sexual abuse, Flannel Flower with Fringed Violet will help men recover, while for women Fringed Violet and Wisteria give wonderful results.

Many women in long-term relationships feel a great deal of frustration because their partners don't communicate with them. For those men who don't share their thoughts and feelings, Flannel Flower can help them trust that it is safe to reveal their inner selves and express more of the gentle, softer side of their nature. If a man is totally cut off from his feelings, Bluebell will help to open his heart. Research has shown that it is not only women who feel frustration at men's inability to share their feelings. After marriage, very few men have any close male friends in whom they can confide. Although they may maintain old friendships from times when there was closeness—from the army, the old football team or from college—these relationships are not as close any more. To some extent this pattern is changing, but it is still very common among men.

Sundew and Red Lily are useful for people who have difficulty with staying in the here and now. These people are rarely present with their partners during lovemaking as they are totally lost in their own fantasies.

The Paw Paw remedy can be used by couples who cannot decide whether or not to have a child, and Illawarra Flame Tree by those who delay parenthood because they feel overwhelmed by the responsibility of rearing children. She Oak can be taken by women who want children but are unable to conceive. Before deciding to have children, couples can take Wedding Bush to commit themselves to their relationships.

Many women possess a very strong water element, in the metaphysical sense of the term. These women consider the needs of their children to be their highest priority. The water female may experience great difficulty in balancing her role as a mother with her role as a wife, finding she is cutting off sexually and emotionally from her partner, devoting her time to her children. Bush Gardenia can help to bring about a balance in these women's lives.

The Kapok Bush remedy is for those who are half-hearted and give up quite easily. Some couples rarely make love because they feel it is too much of an effort. Bush Gardenia as well as Kapok Bush can be used in these situations.

Black-eyed Susan or *Banksia robur* is the essence for ambitious people who burn themselves out, not leaving enough time or energy for their relationships or for sex. For the other extreme, sexual excess, Bush Iris can be used by people who are too preoccupied with the physical.

As individuals age, they pass through major biological transitions such as pregnancy, parenthood, menopause, etc. Bottlebrush will help them cope with these changes, so that they won't feel overwhelmed by them.

In old age, many men experience the frustration of feeling passion but being unable to express it physically. Wild Potato Bush is the remedy for this situation.

As their bodies age, some people lose their self-confidence. These people may want to attract a partner and create a new relationship, yet they hold back because of a lack of confidence in their attractiveness. Five Corners will help to restore their feelings of self-esteem.

When a relationship ends through the death of a partner, Fringed Violet can be taken for the shock, and Sturt Desert Pea for the deep hurt. Other remedies that allow one to cope with the loss of a partner are Boronia for the pining, and, to add some joy to life, Little Flannel Flower.

Bluebell and Flannel Flower are good remedies for keeping a relationship alive and growing over a long period of time.

As you have seen from this very brief overview of sexuality, many bush essences can play a very important role in improving the quality of our lives from conception to old age.

—Meditation—

In earlier sections of this book you will have no doubt noticed the emphasis that has been placed on turning inward and finding the still centre within yourself. There you will find the answers to everything you need to know and you will be spiritually, physically and emotionally recharged. It was from the quiet centre within myself that I was given the information on the Australian Bush Flower Essences and an understanding of them.

Thus this section on meditation is very much in keeping with the goal of this book; namely, to help awaken in you the sense of how magnificent you are and to give you a glimpse of your full potential. The bush essences and meditation offer you the tools to assist you in your realisation of that potential.

There have been many books written on meditation, yet the act of meditating is extremely simple. You do not need to spend thousands of dollars on a special course to learn meditation as it is very easy to learn.

You can, of course, create as many rituals as you like around the act of meditating. Burning a candle during meditation, for example, generates a nice energy in the room. Praying directly before meditating is another excellent way, to evoke a good feeling in yourself as well as a calming, protective vibration around you.

The Indian practice of hatha-yoga is used to relax and settle the physical body and to balance the endocrine glands so that meditation is more effective and much deeper. So before meditating, if possible, have a good stretch and loosen up with some exercises.

People with a busy schedule—those who would obtain excellent benefits from meditating—often have difficulty in finding the time to meditate each day. The best solution to this problem is to make meditation a top priority each day and to schedule everything else around it. It is indeed a sad indictment of urban living that putting aside twenty minutes for oneself can be so difficult.

It is preferable to meditate at about the same time each day. However, it is far better to meditate on a daily basis than to meditate a couple of times a week at a particular time. Meditating within an hour and a half of eating can be physically uncomfortable, as there is a feeling of heaviness in the stomach. Sunrise and sunset are often suitable times for meditating, but choose a time that is most convenient for you.

Twenty minutes is the normal length of time for meditating. Research has shown that twenty minutes of meditation is equivalent to two hours of sleep, so there are certainly advantages in making this a daily priority. It is helpful to meditate with a group of people, say, once a week, and to practise on your own on the other six days.

The main purpose of meditation is to still the chatter of the conscious mind in order to increase your perception of your intuitive nature and to allow life's dramas and frustrations to fall away or significantly pale.

First, get yourself into a comfortable position. Your body energies will flow more freely and you will be more comfortable if you sit with your back straight and your arms and legs uncrossed. You can, of course, lie down, but then you are more apt to fall asleep.

A number of different techniques can be used. A Zen method is to count the outbreaths: breathe in, and as you breathe out count "one"; on the next outbreath count "two"; continue in this way until you reach "ten"; then repeat. If you become aware that your thoughts have drifted away and that you have lost track of the counting, then gently let go of those thoughts and start again "one". You could consider your thoughts as clouds. As you become aware of them, don't immediately try to squash them but rather acknowledge them and then let them simply drift from your awareness. Don't be too perturbed if you often fail to reach "ten".

Another method of meditation uses a mantra. A mantra is merely a collection of syllables which is rhythmically and silently repeated. It may or may not have a religious connotation. An individual chooses a mantra he or she feels drawn to. Some more popular examples are:

Ah Sum

Jesus Christ

Christ-Um

Om (pronounced Ohm)

Like Zen counting, a mantra gives the mind something to focus on in order to still its chatter. By the end of your meditation you may have totally forgotten about the counting or mantra and may have been "lost" in deep meditation.

Having practised both of the above techniques, I now prefer to use the following method, which is even simpler. I attempt to keep a clear, blank mind throughout my meditation, and if I become aware of any thoughts, then I gently let them slip away.

An important attitude to cultivate towards meditation is to let go of all expectations. One night you may experience a very deep meditation, with the twenty minutes seeming like the blinking of an eyelid. Yet another night you may feel very restless, and the time may seem to drag on. Meditation appears to be more difficult at those times. The temptation is to think of deep meditation as the ideal state, and to believe that you are not really meditating unless you reach that state. However, your Higher Self, or superconscious mind, knows exactly what you need, so simply accept whatever happens in meditation as having been chosen by you, in your best interests. The only thing you need to do is to create the time and space for meditation. The restless, apparently more superficial meditations can be seen as a sorting and clearing out of thoughts from the conscious mind and are as beneficial as profound meditations.

After a little practice you will quickly become aware of when the twenty minutes are up, but in the interim it is quite all right to come out of your meditation and have a look at a watch or a clock. If you have not been meditating for the full twenty minutes, just go back into meditation for the remaining time.

Occasionally you may experience a lot of twitching or fidgeting in your feet or hands, and some may even feel pins and needles in their legs. Don't be alarmed, as these are signs that energistic blocks in the body are being cleared during meditation. In a deep meditation your head may sometimes nod forward and then jolt back. Again, don't worry; this is often caused by the clearing of blocks around the throat area and can even be an indication that your spirit guides are working in this area and that you have mediumistic abilities.

You may find that your sittings are very productive times in which problems are easily solved and great insight and understanding comes through to you. With twenty minutes of meditation daily, you may save yourself untold hours of worry and effort and you will greatly improve the quality of your life.

The Australian Flower Remedy Society

The Society has been formed to provide you with the most up-to-date information on the Australian Bush Flower Essences, and, at this stage, there is a $10 annual membership fee. Members regularly receive four newsletters a year which contain updates on the remedies, details of future workshops and information about discounts on products. They also provide a forum, allowing members to share their knowledge of and experiences with the remedies.

To become a member, simply contact the Australian Flower Remedy Society and leave your name, address and telephone number.

Supply

For information regarding the supply of remedies, please contact the Australian Flower Remedy Society, PO Box 531, Spit Junction, NSW 2088, Australia.

—Bibliography—

Bach, Edward. *Heal Thyself*. C.W. Daniel Co., Essex, 1931.
——. *Twelve Healers and Other Remedies*. C.W. Daniel Co., Essex, 1933.

Bach, Richard. *A Bridge Across Forever*. Pan Books Ltd, London, 1985.

Baker, Douglas. *Esoteric Healing*. Pt III, Douglas Baker, England, 1978.

Baker, Margaret, Corringham, R. & Dark, J. *Native Plants of the Sydney Region*. Three Sisters Productions, Sydney, 1986.

Bellin, Gita. *Amazing Grace*. Gita Bellin Associates, Sydney, 1987.
—— Self Transformation Centre, The. *A Sharing of Completion and Celebration*. Self Transformation Seminars Ltd, Sydney, 1983.

Blombery, A. *What Wild Flower Is That?* Lansdowne Press, Sydney, 1973.

Bolton, B.L. *The Secret Power of Plants*. Abacus, London, 1975.

Boone, J. Allen. *Kinship with All Life*. Harper and Rowe, New York, 1954.

Brennan, Kym. *Wild Flowers of Kakadu*. K. G. Brennan, Jabiru, 1986.

Brilliant, Ashleigh. *I have Abandoned My Search for Truth and am Now Looking for a Good Fantasy*. Woodbridge Press Publishing Co., California, 1980.

Brock, John. *Top End Native Plants*. John Brock, Darwin, 1988.

Burnum Burnum. *Burnum Burnum's Aboriginal Australia*. Angus & Robertson Publishers, Sydney, 1988.

Caddy, Eileen. *The Dawn of Change*. Findhorn Press, Findhorn, Scotland, 1979.
——. *Footprints on the Path*. Findhorn Press, Findhorn, Scotland, 1976.
——. *God Spoke to Me*. Findhorn Press, Findhorn, Scotland, 1971.
——. *Opening Doors Within*. Findhorn Press, Findhorn, Scotland, 1986.

Callahan, Dr Roger. *Five Minute Phobia Cure*. Enterprise Publishing Inc., Wilmington, USA, 1985.

Collins, Tom. *Such Is Life*. Angus & Robertson Publishers, Sydney, 1944.

Conabere, Elizabeth & Garnet, J. R. *Wild Flowers of South Eastern Australia*. Greenhouse, Melbourne, 1987.

Douglas, Nick & Slinger, Penny. *Sexual Secrets*. Destiny Books, New York, 1979.

Druck, Ken. *The Secrets Men Keep*. Doubleday & Co. Inc., 1985.

Erickson, Rica. *Orchids of the West*. University of WA Press, Perth, 1965.
——, George, A. S., Marchant, N. G. & Morcombe, M. K. *Flowers and Plants of Western Australia*. Reed, Sydney, 1973.

Fritz, Robert. *The Path of Least Resistance*. DMA, Salem, 1984.

Gardener, C. A. *Wildflowers of Western Australia*. St George Books, Perth, 1959.

Gawain, Shakti. *Creative Visualisation*. Bantam Books Inc., New York, 1979.

Gerber, Richard, MD. *Vibrational Medicine*. Bear & Co., Sante Fe, 1988.

Gibbs, May. *Flannel Flower Babies*. Angus & Robertson Publishers, Sydney, 1983.

Gibson, Jack. *Played Strong, Done Fine*. Lester-Townsend Pub., Sydney, 1988.

Gledhill, D. *The Names of Plants*. Cambridge University Press, London, 1985.

Greenaway, Kate. *The Illuminated Language of Flowers*. Macdonald and Janes, London, 1978.

Gurudas. *Flower Essences*. Brotherhood of Life Inc., New Mexico, 1983.

Hay, Louise L. *Heal Your Body*. Specialist Publications, Sydney, 1976.

Hayward, Susan (ed.). *Begin It Now*. In-Tune Books, Sydney, 1987.
——. *A Guide for the Advanced Soul*. In-Tune Books, Sydney, 1985.
—— & Cohan, Malcom (eds). *A Bag of Jewels*. In-Tune Books, Sydney, 1988.

Huxley, Anthony. *Green Inheritance*. William Collins Sons & Co. Ltd, London, 1984.
——. *Plant and Planet*. Allen Lane, London, 1974.

Jampolsky, Gerald G. *Love Is Letting Go of Fear*. Celestial Arts, California, 1979.

Langloh Parker, K. *Australian Legendary Tales*. The Bodley Head Ltd, London, 1978.

Maclean, Dorothy. *To Hear the Angels Sing*. Lorian Press, Washington, 1980.

Maple, Eric. *The Secret Lore of Plants and Flowers*. Robert Hale Ltd, London, 1980.

Molyneux, Bill. *Bush Journeys*. Thomas Nelson, Melbourne, 1985.

Mountford, Charles T. *Winbraku and the Myth of Jarapiri*. Rigby, Adelaide, 1968.

Neidjie, Bill. *Australia's Kakadu Man*. Mybrood, 1985.

Nixon, Paul. *The Waratah*. Kangaroo Press, Sydney, 1987.

O'Connor, Dagmar. *How to Make Love to the Same Person for the Rest of Your Life*. Columbus, London, 1985.

Odent, Michel. *Birth Reborn*. Pantheon Books, New York, 1984.

Pearce, Joseph Chilton. *Magical Child Matures*. Bantam Books, Toronto, 1985.

Pepper, Frank S. (ed.). *20th Century Quotations*. Sphere Books Ltd, London, 1984.

Phillips, David A. *Secrets of the Inner Self*. Angus & Robertson Publishers, Sydney, 1980.

Proctor, John & Susan. *Nature's Use of Colour in Plants and Their Flowers*. Cassell, 1978.

Rajneesh, Bhagwan Shree. *Dying for Enlightenment*. Rajneesh Foundation International, 1979.

Ramtha. *Ramtha* (ed. S. L. Weinburg). Sovereigny Inc., Washington, 1986.

Ray, Sondra. *Ideal Birth*. Celestial Arts, Berkeley, 1985.
——. *Loving Relationships*. Celestial Arts, Berkeley, 1980.

Rentoul, J. N. *Growing Orchards*. Lothian Publishing Co., 1985.

Rintoul, Stuart. *Ashes of Vietnam*. William Heinemann, London, 1987.

Robbins, Anthony. *Unlimited Power*. Simon and Schuster, New York, 1986.

Roberts, Jane. *The Nature of Personal Reality*. Prentice-Hall, New Jersey, 1974.

Serventy, Vincent. *Australian Native Plants*. Reed, Sydney, 1984.
——. *Plant Life of Australia*. Cassell, Sydney, 1981.

Stair, Nadine. "If I Had My Life to Live Over" in *Chop Wood, Carry Water* (ed. New Age Journal). Jeremy P. Tarcher, Los Angeles, 1984.

Steven, Margaret. *First Impressions: The British Discovery of Australia*. British Museum, London, 1988.

Thie, John. *Touch for Health*. De Vorss and Co., Marina del Rey, California, 1973.

Vaughan, Frances & Walsh, Roger (eds.). *A Course in Miracles*. Jeremy P. Tarcher, Los Angeles, 1983.

Verny, T. *The Secret Life of the Unborn Child*. Sphere Books, London, 1981.

Vlamis, G. *Flowers to the Rescue*. Thorsons Publishing Group, Wellingborough, 1986.

Weeks, Nora. *The Medical Discoveries of Edward Bach, Physician*. Keats Publishing Inc., New Canaan, 1979.

Wylie, Philip. *Generation of Vipers*. Muller, London, 1955.

Index
of Illnesses and
– Their Treatment –

Throughout this book much emphasis has been placed on the connection between emotional patterns and imbalances and physical illness. The purpose of this index is to give further insight into emotional patterns and other factors that can lead to disease. The index contains a listing of illnesses and specific anatomical features. The index is to be used in the following way: find the illness, note the essence or essences recommended in treatment and refer to the description of the essence in the text. For example, under the heading "arthritis" the following essences are listed: Southern Cross, Yellow Cowslip Orchid, Dagger Hakea, Hibbertia, Little Flannel Flower, Isopogon. Anyone suffering from arthritis can read the material on each of those six essences, decide which is most suitable and work with that essence or essences to help resolve that pattern. In no way should this index be seen as a claim that the essences will cure the disease. They should never replace the services of a qualified practitioner.

abscess
    Dagger Hakea
    Mountain Devil

**accidents, treatment of
trauma in**
    Boronia
    Jacaranda
    Sundew
    Red Lily
    Mountain Devil
    Fringed Violet

aches
    Five Corners
    Bluebell
    Tall Yellow Top

acne
    Spinifex
    Five Corners
    Billy Goat Plum

addictions
    Bottlebrush
    Waratah
    Red Lily
    Sundew
    Flannel Flower
    Five Corners

adenoids
    Red Helmet
    Dagger Hakea

adrenals
    Macrocarpa
    Black-eyed Susan

ageing, retardation of
    Peach-flowered Tea-tree
    Mountain Devil
    Bauhinia
    Dagger Hakea
    Little Flannel Flower

AIDS
    Sturt Desert Rose
    Waratah
    Illawarra Flame Tree

allergies
    Fringed Violet
    Dagger Hakea

amenorrhoea
    Five Corners
    She Oak

amnesia	Isopogon Sundew Red Lily Little Flannel Flower
anaemia	Five Corners Kapok Bush Bluebell
ankles, problems of	Flannel Flower Sturt Desert Rose Isopogon
anorexia	Five Corners Dagger Hakea Grey Spider Flower
anus, disorders of	Bottlebrush Sturt Desert Rose
anxiety	Dog Rose Crowea
appetite, disorders of	Dog Rose Paw Paw Crowea Five Corners Bluebell
arms, pain in	Paw Paw
arteries	Bluebell
arteriosclerosis	Isopogon Bottlebrush Yellow Cowslip Orchid
arthritis	Southern Cross Yellow Cowslip Orchid Dagger Hakea Hibbertia Little Flannel Flower Isopogon Sturt Desert Pea
asthma	Bluebell Red Grevillea Tall Yellow Top Grey Spider Flower

aura, broken	Fringed Violet
aura, misaligned	Crowea
back, general	Waratah
	Paw Paw
	Sunshine Wattle
lumbar and sacral region	Crowea
	Southern Cross
thoracic region	Crowea
	Sturt Desert Rose
	Bottlebrush
cervical vertebrae	Bluebell
	Paw Paw
	Tall Yellow Top
balance	Jacaranda
	Bush Fuchsia
	Crowea
baldness	Dog Rose
	Hibbertia
	Yellow Cowslip Orchid
bedwetting	Red Helmet
	Dog Rose
bites, insect	Mountain Devil
blisters	Spinifex
	Fringed Violet
blood disorders	Bluebell
	Bottlebrush
blood pressure, high	Crowea
	Five Corners
	Mountain Devil
blood pressure, low	Southern Cross
	Kapok Bush
	Five Corners
body odour	Dog Rose
	Five Corners
	Billy Goat Plum

boils	Mountain Devil
bone fracture	Fringed Violet Red Helmet
bone marrow	Five Corners
brain imbalances	Bush Fuchsia Sundew Isopogon
breasts	Philotheca *Banksia robur* Bottlebrush
breathing problems	Sunshine Wattle Five Corners Tall Yellow Top
bronchitis	Dagger Hakea
bruising	Five Corners Flannel Flower
bulimia	Billy Goat Plum Grey Spider Flower Five Corners
burns	Mulla Mulla Dagger Hakea Mountain Devil
callouses	Yellow Cowslip Orchid Bauhinia
cancer	Sturt Desert Pea Dagger Hakea Mountain Devil Slender Rice Kapok
candida	Spinifex Kangaroo Paw
car sickness	Red Grevillea Dog Rose
cataracts	Sunshine Wattle Waratah

cellulite	Dagger Hakea
	Bottlebrush
	Billy Goat Plum
cheeks	Sturt Desert Rose
chin	Tall Yellow Top
	Five Corners
chlamydia	Spinifex
cholesterol imbalance	Black-eyed Susan
	Bluebell
	Flannel Flower
	Little Flannel Flower
chronic disease	Kapok Bush
	Bauhinia
	Sunshine Wattle
	Dog Rose
circulation problems	Bluebell
	Flannel Flower
colds	Paw Paw
	Jacaranda
	Black-eyed Susan
colic	Black-eyed Susan
	Paw Paw
	Emergency Essence
colitis	Bottlebrush
	Dog Rose
coma	Sundew
	Red Lily
	Emergency Essence
conjunctivitis	Mountain Devil
	Sunshine Wattle
constipation	Bluebell
	Bottlebrush and Boronia
corns	Isopogon
	Bauhinia

coughs	Illawarra Flame Tree Dagger Hakea Red Helmet
cramps, muscle	Grey Spider Flower Bottlebrush
cuts	Spinifex Sturt Desert Rose
cystic fibrosis	Southern Cross
cystitis	Dagger Hakea Bottlebrush
cysts	Sturt Desert Rose Mountain Devil
deafness	Tall Yellow Top Isopogon Illawarra Flame Tree
dehydration	She Oak
depression	Kapok Bush Flannel Flower Wild Potato Bush Red Grevillea
diarrhoea	Paw Paw
dizziness	Jacaranda Bush Fuchsia
dwarfism	Yellow Cowslip Orchid
dyslexia	Bush Fuchsia Sundew Jacaranda
ears, problems of	Kangaroo Paw Bush Gardenia
eczema	Dagger Hakea Billy Goat Plum
elbow	Bottlebrush Bauhinia

Epstein-Barr virus

Five Corners
Macrocarpa
Banksia robur
Bottlebrush

eyebrow

Dagger Hakea
Mountain Devil

eyes

Sunshine Wattle
Bush Fuchsia

Fallopian tubes

She Oak
Spinifex

fatigue

Macrocarpa
Sunshine Wattle
Old Man Banksia
Banksia robur

fever

Mountain Devil
Mulla Mulla

fingers, pain in

Sundew
Kapok Bush

food poisoning

Paw Paw
Crowea

foot problems

Sunshine Wattle
Bauhinia
Dog Rose
Silver Princess
Bottlebrush

forehead

Paw Paw
Boronia
Crowea

frigidity

Billy Goat Plum
Wisteria
Red Helmet
Flannel Flower

frustration

Banksia robur
Old Man Banksia
Black-eyed Susan
Jacaranda
Sunshine Wattle
Red Grevillea

gallstones
Dagger Hakea
Southern Cross
Slender Rice Flower

gout
Mountain Devil
Black-eyed Susan

gum problems
Jacaranda
Peach-flowered Tea-tree

haemorrhoids
Bottlebrush
Mountain Devil
Black-eyed Susan

halitosis (bad breath)
Mountain Devil
Paw Paw
Crowea

headaches
Emergency Essence
Five Corners
Paw Paw
Sturt Desert Rose

heart
Bluebell
Black-eyed Susan
Red Helmet
Old Man Banksia
Little Flannel Flower

herpes
Sturt Desert Rose
Billy Goat Plum
Spinifex

hiccups
Paw Paw
Black-eyed Susan
Crowea

hip, problems of
Sundew
Sunshine Wattle
Dog Rose
Old Man Banksia

hives
Dog Rose
Fringed Violet
Dagger Hakea

hyperactivity
Black-eyed Susan

hypoglycaemia	Paw Paw Peach-flowered Tea-tree Kapok Bush
hypothalamus	Bush Fuchsia
impotence	Crowea Flannel Flower Five Corners
incontinence	Hibbertia Bottlebrush Five Corners
indigestion	Dog Rose Paw Paw Crowea Black-eyed Susan
infection, bacterial	Black-eyed Susan Mountain Devil Dagger Hakea
infertility	She Oak Turkey Bush
influenza	Paw Paw Black-eyed Susan Jacaranda
insanity	Waratah Tall Yellow Top Sundew
insomnia	Black-eyed Susan Crowea Boronia
itching	Red Grevillea Black-eyed Susan
jaw, problems of	Mountain Devil Dagger Hakea
jet lag	Crowea+ Bottlebrush+ Paw Paw+ Sundew+

	Banksia robur+ She Oak+ Wild Potato Bush (combination)
kidney	Dog Rose Red Grevillea Grey Spider Flower
knee, problems of	Isopogon Macrocarpa Wild Potato Bush
laryngitis	Red Helmet Bush Fuchsia
liver, disorders of	Mountain Devil Dagger Hakea Southern Cross Slender Rice Flower
lung problems	Sturt Desert Pea Boronia
malaria	Paw Paw
memory, failure of	Isopogon
menopause	Bottlebrush She Oak Ilawarra Flame Tree Peach-flowered Tea-tree
menstrual problems	Billy Goat Plum She Oak Sturt Desert Rose
mouth, problems of	Bauhinia Isopogon Billy Goat Plum
multiple sclerosis	Isopogon Bluebell Hibbertia
muscles	Black-eyed Susan Wild Potato Bush Bauhinia

nails	Fringed Violet
nausea	Dog Rose Paw Paw
neck	Isopogon Crowea Kangaroo Paw
nervous breakdown	Black-eyed Susan Old Man Banksia Fringed Violet Jacaranda Paw Paw
nervousness	Dog Rose Black-eyed Susan Jacaranda
neuralgia	Sturt Desert Rose Flannel Flower
nosebleeds	Illawarra Flame Tree
nose, running	Flannel Flower
ovaries	She Oak Turkey Bush
overweight	Old Man Banksia Wild Potato Bush Mulla Mulla Dog Rose Fringed Violet
pain, general	Sturt Desert Rose Emergency Essence
pancreas	Peach-flowered Tea-tree
paralysis	Wild Potato Bush Grey Spider Flower
parasites	Southern Cross Billy Goat Plum Kapok Bush Five Corners

peptic ulcer	Crowea
pimples	Five Corners Billy Goat Plum
Pineal gland	Bush Fuchsia
pituitary gland	Yellow Cowslip Orchid
premenstrual tension	She Oak Crowea Peach-flowered Tea-tree
prostate	Flannel Flower Sturt Desert Rose Kapok Bush
psoriasis	Little Flannel Flower Billy Goat Plum
radiation	Fringed Violet + Bush Fuchsia + Paw Paw + Waratah + Crowea + Mulla Mulla (combination)
rape, trauma of	Wisteria + Fringed Violet Flannel Flower + Fringed Violet (combination) Billy Goat Plum
rheumatism	Isopogon Yellow Cowslip Orchid Southern Cross Dagger Hakea
rheumatoid arthritis	Dagger Hakea Hibbertia Sturt Desert Pea Red Helmet Southern Cross
round shoulders	Waratah Dog Rose Five Corners Sunshine Wattle
RSI (repetitive strain injury)	Southern Cross Red Grevillea

sciatica

Dog Rose
Crowea

senility

Isopogon
Sundew
Red Lily

shock

Emergency Essence
Fringed Violet

shoulders

Paw Paw
Dog Rose
Sunshine Wattle

sinusitis

Dagger Hakea

skin problems

Five Corners
Billy Goat Plum
Bottlebrush
Fringed Violet
Jacaranda

smoking, difficulty in quitting

Boronia + Bottlebrush

snoring

Old Man Banksia
Isopogon

spleen

Dagger Hakea
Boronia
Hibbertia

stiffness

Yellow Cowslip Orchid
Bauhinia
Isopogon
Little Flannel Flower
Hibbertia

stomach problems

Paw Paw
Crowea
Dog Rose
Billy Goat Plum

stroke

Emergency Essence
Old Man Banksia
Black-eyed Susan
Kapok Bush
Bauhinia
Tall Yellow Top

vaginitis   Dagger Hakea
                                  Sturt Desert Rose
                                  Billy Goat Plum

varicose veins   Paw Paw
                                  Red Grevillea
                                  Banksia robur

venereal disease   Dagger Hakea
                                  Sturt Desert Rose
                                  Billy Goat Plum

vomiting   Bauhinia
                                  Paw Paw

warts   Five Corners
                                  Billy Goat Plum